David Page

The Past and Present Life of the Globe

Being a Sketch in Outline of the World's Lifesystem

David Page

The Past and Present Life of the Globe
Being a Sketch in Outline of the World's Lifesystem

ISBN/EAN: 9783744712071

Printed in Europe, USA, Canada, Australia, Japan

Cover: Foto ©ninafisch / pixelio.de

More available books at **www.hansebooks.com**

THE WORLD'S LIFE-SYSTEM

"The investigation of nature teaches us to recognise the omnipotence, the perfection, and the inscrutable wisdom of an infinitely higher Being, in his works and actions So long as we are ignorant of these things, the perfect development of the human mind cannot be hoped for, or even conceived. Without this knowledge the immortal spirit of man cannot attain to a consciousness of its own dignity, or of the rank which it occupies in creation."—LIEBIG'S *Familiar Letters on Chemistry.*

THE

PAST AND PRESENT LIFE

OF

THE GLOBE

BEING A SKETCH IN OUTLINE

OF

THE WORLD'S LIFE-SYSTEM

BY

DAVID PAGE, F.G.S.

AUTHOR OF "INTRODUCTORY AND ADVANCED TEXT-BOOKS OF GEOLOGY"
"HANDBOOK OF GEOLOGICAL TERMS," ETC.

WILLIAM BLACKWOOD AND SONS
EDINBURGH AND LONDON
MDCCCLXI

WILLIAM MILLER, ESQ.

IN PLEASANT REMEMBRANCE

OF HOURS AND EVENINGS OF DISCUSSION:

OUR THEME—

THE THEME OF THE FOLLOWING PAGES.

PREFACE.

THE object of the following Chapters is to present a sketch in outline of the World's Life-System—tracing from the earliest organisms in the stratified crust to the forms that now adorn and people its surface. The aim has been to link the remote to the recent—the living to the extinct—that the general reader may be enabled to form some intelligible conception of the whole as a great and continuously-evolving scheme of vegetable and animal existences. There is no attempt whatever to teach anatomical details or point out specific distinctions—the volume being intended not as a Handbook of Palæontology, but simply as a readable sketch for the information of those who have neither the time nor the preliminary training to avail themselves of works of higher scientific pretensions. And yet the reader will find in these pages a reliable *résumé* of the science, as founded on the most recent discoveries, and a treatment of its bearings from a higher stand-point than can be conveniently taken by the mere text-books and manuals of Geology. At a time when the question of Life is receiving a wider audience, such a *résumé* may also be of utility in indicating the line that separates the assumed and hypothetical from the known and ascertainable; and so prevent the unprofessional inquirer from ascribing to Geology what it does not affirm, or from expecting from its teachings what they cannot reveal.

Designed for the general reader, and delivered in part to popular audiences, the style is, perhaps, somewhat more rhetorical than befits the exactitudes of science; but even on this point the Author could not well have done otherwise. His object was to excite rather than satisfy the curiosity of his hearers—to impress them with the universality and uniformity of natural law—believing there can be no true notion of Nature or of Nature's requirements while her facts are viewed through the medium of the miraculous. Nor let it be thought that, by recognising in every instance the fixity and unerring operation of Law, we place a wider distance between the Creator and his works, or that any knowledge of this kind has a tendency to self-sufficiency or irreverence. On the contrary, he who knows most of creational law, and that the most intimately, stands generally the least in need of the injunction—"Put off thy shoes from off thy feet, for the place whereon thou treadest is holy."

In treating such a theme as Life—its apparent origin and progress—the writer has necessarily had occasion to allude to subjects on which there is much diversity of opinion; to some that are usually approached with uneasy tenderness, as coming in conflict with prevalent beliefs; and to others on which the united labours of Geologists, during the last fifty years, have thrown but little reliable light or information. In either case he has expressed his opinions freely, but without dogmatism; firmly, but solely under the warrant of Geology; and always with a frank admission of the many deficiencies and imperfections of that science. As there is nothing to be gained by offending a prejudice where we cannot establish a conviction, he has contented himself by stating what Geology affirms, without alluding to what it appears to contradict; and as the establishment of truth does not always follow the overturning of error, the expounder of science may surely be permitted to attempt the one without hazarding an endeavour to accomplish the other. In approaching our subject, therefore,—a subject too often treated as if it lay beyond the pale of natural law,—let it be clearly understood that we are dealing with Life solely in its geological aspects. We appeal unto Cæsar; let us be judged by Cæsar's laws.

GILMORE PLACE, EDINBURGH,
    February 1861.

CONTENTS.

INTRODUCTORY.

                    PAGE

Interest attached to the study of the PAST in natural as in human history.—Fossils, or petrified remains of plants and animals, the medals and records of Creation.—The unerring certainty of the record.—Palæontology, or the Science of Extinct Life.—Its scope and bearings, as founded on a knowledge of the present life of the globe.—Its importance, abstract and practical.—The task it has yet to accomplish, ... 17

THE PRESENT.

Characteristics and classification of living plants and animals as established by the Botanist and Zoologist: 1. Plant-Life.—Its governing conditions in space.—Its typical forms and characters.—Its primal plan and patterns.—Systematic arrangement of its forms.—Their apparent functions.—Persistency of plan in time past. 2. Animal Life.—Its distribution or governing conditions in space.—Its typical forms and their functions.—Its primal plan and patterns.—Higher and lower.—Systematic arrangement of its forms.—Identity of plan and design in time past. 3. Co-adaptation of plants and animals in one great Life-Scheme, ... 27

THE RECORD.

Chronology of geology, or the arrangement of the world's past into Rock-formations and Life-periods.—Principles and methods of this arrangement. — Continuity of natural law. — Provisional and negative state of geological knowledge as influencing our comprehension of the successional order of organic being.—Palæontology so based.—The problems it has to solve.—Its progress and prospects, ... 69

THE FAR PAST.

Characteristics and gradations of the PALÆOZOIC or "Ancient Life" period: 1. The Cambrian age—so-called "Dawn of Life." 2. The Silurian age.—Erroneous notions respecting its physical geography and life-relations.—Its vegetation, graptolites, corals, star-fishes, shell-fish, and crustaceans.—Their specialties and place in the scale of being. 3. The Devonian or Old Red Sandstone age.—Physical features of the epoch.—Its plants, crustaceans, shell-fish, and fishes. 4. The Carboniferous age.—The physical geography and climatal conditions of the period.—Its forest-growths, coral-reefs, shell-beds, crustaceans, insects, fishes, and reptiles. 5. The Permian or Lower New Red Sandstone age—so-called "close" of the Palæozoic cycle. Imperfect interpretation and provisional nature of the Life-phases and Life-periods of the geologist, ... 79

THE MIDDLE PAST.

Characteristics and gradations of the MESOZOIC or "Middle Life" period: 6. The Triassic or Upper New Red Sandstone age.—Its foot-tracks, birds, and reptiles. 7. The Oolitic age.—Sea and land of the epoch.—Its vegetation, lower marine life, shell-fish, crustacea, insects, reptiles, and terrestrial mammals. 8. The Cretaceous or Chalk age.—Physical geography of its seas and shores.—Its lower marine life, foraminiferæ, sponges, star-fishes, sea-urchins, shell-fish, fishes, reptiles, birds, and mammals.—Generalisations resulting from a review of the Mesozoic cycle, ... 119

THE RECENT.

Characteristics and gradations of the CAINOZOIC or "Recent Life" period: 9. The Tertiary age.—Geography of the epoch.—Its huge terrestrial mammals and recent forms of life.—Intermediate forms, and their relation to the fauna of existing areas.—Extinctions during the so-called "Glacial" or "Drifts" period. 10. The Post-Tertiary or Current age.—General existing arrangements of sea and land.—Existing forms and distribution of life.—General and local extinctions.—Man, pre-historic and historic.—Mutations of the human race, . . . 151

THE LAW.

PAGE

General deductions arising from the discoveries of Palæontology.—Origin and advent of life unknown to science.—Universality of life in time and space.—Uniformity of type and plan through all the geological life-periods.—Similarity of functions and life-relations.— Distribution in space.—External conditions never uniform. — Representatives of the great life-types in every epoch. — Gradation and progress.—The course and apparent order of this progress.—Introduction of new forms.—Extinction and creation of species.—Theories of variation and development.—Geological epoch of man.— Time and progress. — Apparent course of creation.—Life-phases of the Current epoch.—Causes of local and general extinction.—Man as a sub-creative centre and modifying agent.—Duration of species.—Time and term of the human race.—Life aspects of the future.—Progression or succession?—Onward and upward, ... 177

CONCLUSION.

What has been aimed at.—The known and the unknown.—The field in which we may all become fellow-workers.—The spirit in which we should inquire, ... 241

INTRODUCTORY

PALEONTOLOGY—ITS SCOPE AND BEARINGS

THESE fragments of rock on the table before us, chips which the road-maker would consider sorry material for his purpose, and the feet of the ignorant might spurn from their path, are in the eye of Science invested with as high an interest as the obelisks of Egypt, or the sculptures of Nineveh. The antiquarian pores with enthusiasm over the lines and letters of the one, and endeavours to decipher the unconnected history of a few thousand years; the geologist bends with equal delight over the forms and impressions on the other, and tries to gather therefrom some intelligible glimpses of a PAST, compared with whose duration the chronology of man is but as the moments of yesterday. The one as connected with the Humanity to which we belong—chequered and humiliating as it has been in many of its phases—must ever excite a lively and immediate interest; the other appertaining to the history of the globe we inherit, and of whose plan our race forms so important a feature, can never cease to attract the attention of enlightened intelligence. In his inciting research the archæologist exhumes buried cities and catacombs, collects the mutilated fragments of human art, deciphers monumental inscriptions, and notes every vestige of the various races that may have peopled any given locality; so in geology the earnest inquirer examines every accessible stratum, collects the fossil fragments he exhumes, and, comparing them with the plants and animals now peopling the earth, endeavours to arrive at a knowledge of the various races that have successively adorned its surface. As a stone-hatchet, a flint arrow-head, a tree canoe, or fragment of pottery, will often throw a flood of light on the researches of the historian; so in geology, the impression of a leaf, a petrified shell, a tooth, a fragment of bone, or a single fish-scale, will often suffice to unriddle the most puzzling problem. The one kind of evidence speaks of the hand that fabricated, the degree of intelligence that directed the fabrication, and the purpose it was meant to subserve; the other tells of the nature of the plant or animal to which it belonged, the climate and conditions under which it grew and flourished, the place it held, and the function it performed in the world's economy, and, higher than all, the omniscience and skill that pre-ordained and directed with unerring precision its numerous and complicated co-adaptations.

Rough and mutilated as these fragments may appear—obscure as are the forms impressed on their surfaces, they embody a tale of the world's PAST as legible to the eye of Science—and often far more connected—than these sculptures on this slab, or those hieroglyphics graven on that sarcophagus. These forked lobes, little more than a mere discoloration on the stone, once floated as sea-weed in the waters; that reed-like stem converted into stone, as it now is, luxuriated in some primeval marsh; that rock-impressed fern-frond once waved its feathery leaflets in the sunshine of a genial climate; and that tiny spikelet, now the merest film of carbonaceous matter, has sparkled with the night-dews of heaven as certainly as the dews now cherish the tender herb, or the sunlight gives colour to existing verdure. Worthless as these chips may seem, the eye of the zoologist detects in this the pore-work of a coral, in that the valves of a shell-fish; on this the scales of a fish, on that the plates of a reptile; in this the bone of a bird, in that the bone of a mammal; in this the grinder that milled the leafy twigs of the forest, in that the trenchant tooth that preyed on the flesh of other creatures. Every trace becomes a letter, every fragment a word, and every perfect fossil a chapter in the world's history, which tells of waters that were thronged and of lands that were tenanted by life—of races that lived and multiplied and perished—of others that took their places—and this (as we shall afterwards see) so often repeated, over and over and over again, that the mind, at first excited by the marvels it unfolds, begins at last to grow weary of the review, and the finite creature loses itself in the contemplation of the works of the Infinite Creator.

The objects through which we arrive at a knowledge of this extinct life are what are familiarly termed "Fossils"—the remains of plants and animals that were entombed in the silt and sediment of former lakes and estuaries and seas, and became *petrified*, or converted into stone, as these sediments solidified into rocky strata. As the autumnal leaf drops into the stream and becomes imbedded in its mud—as the trees of the forest are borne down by the flooded river and are ultimately entangled in the silt of its estuary—as the coral-reef and shell-bed are gradually increasing and growing, as it were, into limestone before our eyes—as the skeletons of animals are drifted by the tide and fall to the sea-bottom, or sink into rivers and marshes, and are thus preserved from further decay—so in all time past have similar agencies been at work: here preserving the broken twig and the fallen forest, there the coral-reef and the littoral shell-bed, and anon the remains of animals that were borne by rivers from the land, or drifted by the waves on the muddy sea-shore. These organisms so preserved and petrified constitute the "fossils" of the geologist, who, treating them apart from the rocks in which they are imbedded, has erected their study into a new science, under the title of PALÆONTOLOGY, or the Study of Ancient Life. Originally differing in nature, being in various degrees of completeness at the time they were imbedded, and, above all, being preserved in different kinds of rock-matter, as shale, and coal, and limestone, and flint, and sandstone, they are now found in different degrees of perfection and distinctness. In some we find the original form and all the parts entire, of others we have a mere hollow cast or mould, of some a simple impression of the external surface, of others we have but scattered traces, and these so obscure that they can be read only by the higher powers of the microscope; while of many we have no other relic save the passing footprint or the slimy trail that was left on the yielding sands of a former sea-shore. In whatever state they may be found, they are taken up by the palæontologist, compared with existing plants and animals, and arranged, as far as their nature will permit, according to the classifications of the botanist and zoologist. To the palæontologist, therefore, we commit these relics of primeval life, and ask of him to tell—Whether they are the same in kind as those that now adorn our fields and people the land and waters; whether they were of a simpler and lowlier kind that gradually rose, as time rolled on, to their present forms; whether they were of tinier or of more gigantic dimensions; or whether they varied according to external conditions—here dwarfing and dying out, and there some newer creations increasing and spreading under conditions that were favourable to their existence? In fine, we ask of him the history of these extinct forms, as we demand from the botanist and zoologist the history of the plants and animals that now flourish around us; and, combining the living with the extinct, and the recent with the remote, the highest aim of our science is to discover the Creative Plan which binds the whole into one unbroken and harmonious life-system.

It is true that many of these fossils are so fragmentary and obscure that they cannot yet be deciphered, and others are so different from anything now existing in the vegetable or animal world that no definite place can be assigned them. It is also true that the science of Palæontology has little more than passed its infancy, and that of the innumerable relics entombed in the rocky strata of different regions only a small proportion can have yet been discovered. Notwithstanding all this, so enthusiastic has been the research, and so attractive the study, that much satisfactory work has been done, and, by the aid of some of the highest minds in Britain, France, Germany, Italy, and America, palæontology has already taken a permanent place on the roll of human knowledge. Under the hand of a Brongniart, a Goëppert, or a Lindley, these stony stems have started anew into life and verdure, and tangled the swampy jungle or waved in the upland forest; under the reconstructing skill of a Cuvier, an Agassiz, or an Owen, these scattered bones have been reunited in intelligible symmetry, and once more repeopled the earth, the air, and the ocean; while under the magic lenses of an Ehrenberg these muds, and marls, and chalks, have become instinct with life, and ancient waters swarm with innumerable forms.

"The dust we tread upon was once alive."

Much as these and many others have done, year after year is still adding largely to our knowledge of the PAST LIFE of the Globe; and the time, it is hoped, is not far distant when Geology shall be enabled to read, through these fossil chips and fragments, the Life-History of the World, with as much, if not with greater, certainty than we can now read the phases of human history itself, as displayed in the successive developments of Ninevites and Egyptians, of Greeks and Romans, of medieval Goths and modern Anglo-Saxons.

Exciting, however, as this history of the world's Past must be, even to minds the most illiterate, it may be fairly questioned at the outset—To whom, and for what purpose, is all this research and ingenuity expended? Is Palæontology a theme merely for the gratification of idle curiosity and ignorant wonder; or has it, like every true science, qualities of sterling value that appeal at once to the intellectual and physical exigencies of Man? Does it bear in any way on the industrial purposes of life; does it present itself in the light of an exalting intellectual exercise; or, combining both these qualities, does it lead to sounder and more ennobling views of our relationship to God and Creation? If it does neither, it is no true science, and stands unworthy to be ranked with the legitimate subjects of intellectual research. Luckily, however, it does all, and recommends itself, as it were, instinctively to the inquiring and reflective mind. Guided by its deductions, the identification of rock formations, which was formerly in a great measure a matter of hap-hazard, is now a certainty. Fifty years ago the miner and engineer had little to direct them in their researches, save the very variable tints of colour, the structure, or other external aspects of rock-masses. Now, however, a fossil branch, a tooth, or a few scattered fish-scales, will enable them to identify with certainty strata in distant localities, and so save years of unnecessary toil and thousands of useless expenditure. There is, for instance, in Britain a red sandstone *beneath*, and a red sandstone *above*, our most valuable coal-fields—so like in many respects, that which is which mere mineral characteristics cannot always determine. Shall we ignorantly dig through the one for that mineral fuel which never lies beneath it;. or shall we, mistaking the other, maintain that it is folly to pierce through its strata? Where the mere mineralogist stands perplexed, the palæontologist proceeds in the confidence of certainty, from the detection of a *Holoptychian* fish-scale which stamps the existence of the Old Red, or the discovery of a tiny *Palæoniscus* which is equally decisive of the New. Exalted as may be the task of solving the physical and vital problems of the globe, the duty of turning to account its mineral and metallic treasures is not less worthy or important. Science acquires fresh power and position when combined with practice; Philosophy new dignity when ministering to Humanity.

Again, a science that opens up so much of the Past, that reveals so many new forms of life and organisation, cannot fail to have an exalting effect as a purely intellectual exercise. The anatomical reasonings—the skill required to reconstruct such scattered fragments—the detection of means to an end—all this, and much more that must readily suggest itself to the thinking mind, cannot fail to stamp Palæontology as one of the highest themes that can engage enlightened intelligence. Nor is the new light which its deductions have thrown on other branches of natural science among the least of its claims to general attention. The revivifying, as it were, of so many extinct forms of existence has given a new significance to the science of Life; and henceforth no view of the vegetable or animal kingdoms can lay claim to a truly scientific character that does not embody the discoveries of the palæontologist. In fact, so inseparably woven into ONE GREAT SYSTEM are all fossil forms with those now existing, that we cannot treat of the one without considering the other; and can never hope to arrive at a knowledge of Creative Law by any method which, however accurate as regards the one, is not equally careful and accurate as regards the other. Furthermore, connected as the whole phases of external nature are into one beautiful COSMOS, the mind that remains in ignorance of their history can form but a very imperfect, if not an altogether erroneous, notion of its own relationship and connection therewith. For, while the scope of human duty is circumscribed by our relations to external nature, by our relations to our fellow-men, and by our relations to God, a knowledge of these relations as manifested in the great scheme of Creation is altogether indispensable. In the eloquent language of our motto—" So long as we are ignorant of these things, the perfect development of the human mind cannot be hoped for or even conceived. Without this knowledge, the immortal spirit of man cannot attain to a consciousness of its own dignity, or of the rank which it occupies in Creation." Still more: if existing nature furnishes the theologian with irrefragable proofs of unity of plan and design throughout Creation—if his conceptions of Deity are enlarged and his reverence increased by the study of these adaptations—much more must they be exalted when he finds the same harmonies of design and the same unity of plan running through untold ages, and spreading and ramifying through forms so numerous and varied that, varied and rife as existing Life may be, it constitutes but the merest fraction of the Life that has been, and of the forms that have passed away.

Such is the nature and scope of Palæontology—a science whose function is to extract from the sandstones, and limestones, and clays of the stratified crust, the petrified remains of plants and animals, and from these remains to reconstruct the forms to which they belonged, so as to arrive at some intelligible conception of the Life that formerly tenanted the land and peopled the waters. These sandstones, and limestones, and clays, in all their various repetitions, are but the sediments of pre-existing lakes and estuaries and seas; and the fossils they imbed will be more or less perfectly preserved, just as they were deposited in the areas where they lived and grew, or were drifted from a distance in detached and scattered fragments—according as they were rapidly enveloped from further decay, or exposed to the wasting influences of the air and water—and, above all, according to the preservative character of the stratum that contains them. Their imperfection, and the difficulty of reading aright their characters, is greatly increased by the fact that they are for the most part the chance findings of the quarryman and miner, and extracted in chips and fragments even more fragmentary than when originally imbedded. Notwithstanding these obstructions, and the hopelessness of ever obtaining in a fossil state the colours and softer parts that give beauty and outline to animal forms —in spite of the fact that the corresponding portions of structures found to-day may not be turned up even for years to come—and in face of the toil and expense which the study unavoidably entails—substantial progress has been made in Palæontology, and these fragmentary remains of Past Life been reconstructed so as to take intelligible rank and position in the great categories of existing Vitality. Founding on the uniformity of natural law and persistency in the main structural characteristics of plants and animals throughout all time, the Palæontologist, strong in his faith and hopeful of the result, proceeds to his arduous task, and resuscitates as it were the Life of former epochs— clothing the land with verdure and beauty, and peopling the waters with their varied and appropriate forms. Lifting the veil from the Past, he displays the terraqueous aspects of the globe at the successive stages of its history; even as now, through the combined labours of the geographer, the botanist, and the zoologist, we are enabled to present a panorama of existing lands and seas with all their exuberant and varied vitality.

THE PRESENT.

ITS FLORA, FAUNA, AND THEIR CO-ADAPTATIONS.

BEFORE we can rightly compare the Past Life, of which these relics give evidence, with that which now peoples the globe, we must glance at the conditions under which plants and animals at present exist, and know something of their nature and the functions they have to perform. We can only reason respecting the Past from our knowledge of the the Present; and the more intimate our acquaintance with the various phases of existing nature, the sounder our deductions relating to those that have long since passed away. We say *the various phases of existing nature*, for the plants and animals that people the surface of any given latitude may differ altogether in character from those entombed in the strata beneath, and the organisms in the several formations below may now find their nearest analogues in the flora and fauna of distant and diversified regions. If we are familiar, however, with the general conditions under which plants and animals now live and flourish, and if we can establish a relationship between those existing and those long since extinct, then we can recall the conditions under which the latter grew and flourished, and map out the geography and climate of the primeval world, as the geographer now maps out the areas of sea and land, and depicts the various races of life—the belts of sterility and exuberance—and the creative centres from which peculiar families have emanated to perform their functions in the great economy of nature.

I.—ITS FLORA OR PLANT-LIFE.

Glancing first at the VEGETABLE WORLD, we perceive that the great regulators of plant-life are heat, light, and moisture. Such is the order of nature now, and such, we are bound to believe, have been the ordainings of creation from the earliest moment that the vegetable cell was evoked into existence. Under the tropics, both individual exuberance and specific variety attain their maximum intensity; in the temperate zones this intensity gradually declines; while in the arctic regions vegetable life dwarfs and diminishes till it ultimately disappears and gives place to utter sterility. As we start from the equator, each great belt—equatorial, tropical, subtropical, warm-temperate, cold-temperate, subarctic, arctic, and polar[*]—presents its own distinctive features; and though the zones of the southern hemisphere may differ in genera and species from those of the northern, there is still in the respective stages a sufficient resemblance of growth, colouring, and inflorescence, to prove that, latitude for latitude, the prime governing influence is essentially solar. As with latitude, which is influenced in the main by light and heat, so with height above the level of the ocean—an advance upwards into the rarer regions of the atmosphere being equivalent, in some measure, to an advance northwards or southwards into the colder latitudes

[*] The equatorial zone extends on both sides of the equator to about 15° of latitude; the tropical from 15° to the tropics; the subtropical from the tropics to 34°; the warmer temperate from 34° to 45°; the colder temperate from 45° to 58°; the subarctic from 58° to the polar circle; the arctic from the polar circle to latitude 72°; and the polar zone from 72° to the poles.

of either pole.* The mountain that has its base waving with the palms and tree-ferns of India, may have its sides clothed with the oaks and pines of Europe, its higher cliffs with the dwarf-willows and mosses of Nova Zembla, while its snowy peaks are as void of life as the ice-bound shores of the arctic circle. Besides these conditions, there are others of site, or locality, or habitat—conditions which require that the weeds of the ocean should differ from the plants of the marsh, the plants of the marsh from the herbage of the open plain, and the verdure of the plain distinct from that of the mountain forest. Nay more: there are some tribes that will flourish only in rich organic mould, others that prefer the shingly surface of the arid desert; some that exist only on calcareous soils, and others unknown beyond the limits of the salt marsh. Wherever the prime conditions of heat, light, and moisture are present, there the vegetable germ manifests itself—here incrusting the naked rock, there mantling the surface of the stagnant pool—now rooting itself in the decay of its own kind, and at times finding a habitat even in the tissues of the animal structure. More than this: every climatic influence, however faint, leaves its impress on vegetable life. A thicker layer is added to the concentric growth of the timber-tree during a genial than during an ungenial summer; the southern slopes of a hill are more verdant and flowery than those of its northern side; some plants luxuriate in the sea-breeze which would be death to others; and the leafiest side of a tree is ever that which is most accessible to the open sunshine. Again: plants that grow in localities marked by sudden extremes of heat and cold are always more variable in stature, habit, and foliage, than those which flourish under the steadier influences of a genial

* The capacity of the atmosphere for heat decreases with its density, and this density decreases from the level of the ocean upwards.

climate ; and thus we can judge of the climate of a newly-discovered country, as well as of the conditions that prevailed and affected plant-life during the deposition of a rock-formation, which took place thousands of ages ago. Still further, and apparently altogether independent of climate : certain families are restricted to certain regions, beyond which, and under the present arrangements of sea and land, they naturally never pass ; and thus it is that the Cape of Good Hope rejoices in its pelargoniums and geraniums ; China in its teas and camelias ; Australia in its eucalypti and casuarinæ ; the Spanish peninsula in its ever-green oaks ; and the pampas of South America in their gigantic thistles and clover, to the almost total exclusion of other species. Descending from family *regions* to the narrower *provinces* of genera and species, we find some limited to a single valley, to a solitary island, or, it may be, to some particular mountain-slope which, as far as science can perceive, enjoys no external influence that is not equally shared by the other slopes that surround it.

Beyond all these distinctions there is the difference of KIND—a difference for which science can assign no reason, save that it has pleased the Creator so to create them. Why, for instance, does the moss differ from the rush, the rush from the reed, the reed from the willow, the willow from the birch, the birch from the pine, or the pine from the palm ? The oak and the ash grow side by side in the same forest, and yet they are, in the language of naturalists, specifically and generically distinct ; the daisy and wild clover spring from the same soil, and interweave their rootlets to form the same turf, and yet they have no feature or quality in common. That these are facts, the eye of the passing observer may readily perceive ; the reason why, man may never know. It is of little avail to talk of the plasticity of the vegetable organism under the force of external conditions, or to tell us that under these influences the one form is but a modification and development of the other. Even could we establish this fact, and determine the order of its occurrence, it would be no solution of the great primal question of diversity, seeing that plant-life is altogether passive, and that external conditions are of themselves utterly impotent without a higher power to sustain, and an intellect to direct and control, the course of their operations. And, after all, it is less the mere matter of diversity than the *plan* which connects this diversity into one harmonious system; less the apparent order which may be learned than the reason thereof, to which human knowledge may never attain. God has thought fit so to evoke the vegetable kingdom—to invest His works with variety and complexity; and to unravel this complexity, to arrange these various plants according to their kind and character, to classify them, in accordance with the divine design, into species and genera and orders, and to learn their functions and relations, is the task of the student of Nature. Proceeding upon this plan, the botanist arranges plants according to the complexity of their organs, attempting to separate the simpler from the more highly organised, and these again one from another according to certain dissimilarities and differences of form and function. His aim is to discover the creative idea that pervades the vegetable kingdom; and the nearer he approaches that conception, the more intelligible and permanent his so-called "systems" of arrangement.

In attempting this arrangement—numerous, varied, and complex as vegetable life may at first sight appear—the botanist has happily a few great fixed principles in nature to guide him. *Type* and *Order* run unswervingly throughout the whole; and though the Creator might easily have constructed each species after its own type, and rendered plants as varied in their individual forms as they are numerically abundant, yet He has thought fit to restrict Himself, as it were, to a few types and models, and, humanly speaking, like a skilful inventor, to produce an almost endless variety from the co-adaptation of a few simple elements, and complexity of design by the elimination of a few primal patterns. As innumerable hues can be produced from a few primitive colours, as endless strains of music flow from the touches of a few simple chords, or as the ideas of all times and nations can be expressed by the combinations of some twenty or thirty letter-sounds; so in the structure of plants and animals every variety of form, every conceivable adaptation of structure, proceeds from the modification of a few elementary forms and types in nature. Without this uniformity of plan and design, the study of nature by man's limited faculties would have been impossible. Bewildered with variety without design, and lost in complexity without order, the human intellect could never have arrived at any true conception of Nature or of Nature's laws; could never have woven those chains of reason wherewith it may be said to have linked earth to heaven, and affiliated the created to the uncreated Creator. But inasmuch as God is a God of law and of order, and clearly meant those laws and orders to be intelligible to our limited comprehensions, so He has considerately narrowed the bounds of creative design, and made creation a theme at once fitted to exercise our reason, and to draw forth our reverence and love. In studying the vegetable world, therefore, the botanist finds that every diversity of form and structure proceeds from the elimination of a simple CELL, and that this cell-growth lies at the basis of all vegetable development. He also finds that the primary structural form into which it is developed is a LEAF or leaf-like organ; and that this leaf-like organ manifests itself variously—rising from a simple aggregation of cells, as in the sea-weeds and lichens, to the more complex fronds of the ferns and club-mosses; from these to the parallel-nerved leaves of the grasses and palms; and from these again to the reticulated and more highly organised venation of the leaves of the flowering shrubs and true timber-trees. He further finds that, while the leaf is produced by the development of the cell, all the other organs of the plant are but modifications of the leaf —that the stem and branches are elaborated from the successive growths of leaves, that the petals of the flower are but modifications of the same organ for a special purpose, that the fruit is but a specialised combination of leaves, and that the seed itself consists of a leaf or leaves folded up and protected for the return of those conditions of heat and moisture necessary to its starting again into life and verdure, to perform the same round of development and reproduction. How this *cell*, or globule of matter, should become vivified—how it should be capable, under certain conditions of heat, light, and moisture, of being reproduced indefinitely into some determinate form as a leaf, and how these leaves or leaf-like organs should be persistently maintained, each in its own distinctive type throughout the great categories of the vegetable kingdom—are problems which science cannot solve. We know, however, the facts and the order of their occurrence. We perceive the expression of a prescient plan; that plan we endeavour to interpret. Everywhere purpose and design are manifest; into the motives of the Designer we may not inquire. The secondary we may discover: to the primary we can only appeal.

Founding on this great principle of cell and leaf development, the botanist traces its elaboration in the different races of plants, and regards those which manifest little more than a repetition of the same parts as of lower organisation than those in which the leaf is metamorphosed into various organs, each organ having a special function to perform in the plant's growth and perfection. The higher, therefore, that a plant is in the scale of being, the more specialised its organisation; that is, instead of all the functions or several of its functions being performed by the same organ, each function is performed by an organ specially devoted to it. It is thus that the fern is regarded higher than the seaweed; the palm higher than the fern; and the oak than the palm. In ranking plants as "higher" and "lower," the botanist by no means asserts that the one is less fitted than the other for its purpose in creation. All that he affirms—and common-sense homologates the affirmation—is, that the lichen, composed of a mere congeries of cells, and increasing by a mere homogeneous development of these cells, is a less highly organised structure than the timber-tree, in which is elaborated a variety of tissues, which is increased by leaf-growth, and whose reproduction is provided for by a complicated process of flowering and fructification. Aware of these distinctions, and knowing the persistency of nature in her modes of operation, we can determine the relative positions not merely of the plants that now adorn the various regions of the earth, but of those that existed during the successive epochs of her bygone history. As a region of shrubs and timber-trees is said to enjoy a higher flora than a region of ferns and clubmosses, so do the reticulated leaves and concentric woody layers, found fossil in a recent rock-system, give indication of a higher physiological value than the parallel-veined leaves and vascular-bundled stems of some earlier formation. It is thus that we arrive, in general terms, at the great truths of vital progress—a leaf, a stem, the disposition of a branch, or the structure of a fruit, affording such evidence to the palæontologist as the flint arrow-head, the bronze spear, and the primitive matchlock, afford to the archæologist and historian.

Proceeding upon such principles as those indicated in the preceding paragraphs, the botanist arranges all vegetables into two grand divisions—the CELLULAR and the VASCULAR : the former embracing those which, like the mushrooms and lichens and sea-weeds, possess no regular vessels, but are composed of a mere congeries of cells or cellular tissue ; the latter comprising those that are composed of various tissues and furnished with various organs of nutrition and reproduction. Again, he subdivides the vascular into the FLOWERLESS, as the mosses, equisetums, and ferns ; and the FLOWERING, which embraces the palms, and lilies, and grasses, the pines and cycads, and all herbs and shrubs, and true timber-trees. In the Flowerless division (Cryptogams or Sporocarps, as they are sometimes termed) the organs of reproduction are not essentially different from the other parts ; that is, they are not apparent—similar cells forming alike the organs of growth and the organs of reproduction. On the other hand, in the Flowering (Phanerogams or Spermocarps) the organs of reproduction are apparent—the seed being enclosed in an embryo in which the rudiments of the future plant are distinguishable. Still subdividing and arranging, he speaks of Dicotyledons, or those whose seeds, like the bean and acorn, are furnished with *two* lobes ; of Monocotyledons, or those like the palms and grasses, which have only *one* seed-lobe ; of Acotyledons, or those like the ferns and fungi, which have *no* lobes, but are propagated by spores, and so termed *Sporocarps* in contradistinction to the *Spermocarps*, or those bearing true seed-fruits. Again, looking at their modes of development, the botanist speaks of Exogens, or plants whose stems increase by *external* layers of annual growth around a central pith—hence the concentric rings of the ash and beech ;

of Endogens, whose increase takes place *from within* by a coalescence of the footstalks of the old leaves, as in the palm; of Acrogens, or those that increase by shooting *from the top*, as the ferns and horsetails, and whose stems are thus generally thicker above than below; and of Amphigens, or those which grow by additions to the *external margin*, and spread, as it were, on every side, as in the sea-weeds and lichens. Founding in this way—*first*, on the different modes of reproduction; *second*, on the aspect of the reproducing organs; *thirdly*, on the primary development; and *fourthly*, on the ultimate development of the plant—the botanist arrives at a scheme of classification which may be briefly expressed as in the annexed tabulation.

It is true, the palæontologist cannot always avail himself of the terms and classification of the botanist, as there occur in the geological formations a number of forms that stand intermediate between existing orders and families, and of which we have now no living representatives. Still, these forms never diverge so widely from any of the existing families but that their affinities can be determined with some degree of certainty; and at all events, even where family alliance fails, they can be readily ranked under the wider categories of orders and sections. It is thus that the subjoined scheme embraces alike the extant and extinct—the latter supplying the links that unite the whole into a still more homogeneous and consistent system:—

ITS FLORA.

VASCULAR.

SPERMOCARPS or PHANEROGAMS
- ANGIOSPERMS
  - EXOGENS — DICOTYLEDONS......Trees, Shrubs, Herbs.
  - ENDOGENS — MONOCOTYLEDONS...Grasses, Sedges, Palms.
- GYMNOSPERMS
  - GYMNOGENS — POLYCOTYLEDONS...Cycads and Conifers.

SPOROCARPS or CRYPTOGAMS
- ANGIOSPORES
  - ACROGENS
    - SPOROGAMS......Clubmosses, Lycopods.
    - THALLOGAMS......Ferns and Horsetails.
    - AXOGAMS......Mosses and Liverworts.

CELLULAR.

- GYMNOSPORES
  - AMPHIGENS
    - HYDROPHYTES......Algæ and Confervæ.
    - AEROPHYTES......Lichens.
    - HYSTEROPHYTES.....Fungi or Mushrooms.

Or, adopting a simpler and more explanatory arrangement, the several grand divisions of the vegetable kingdom may be exhibited as under :—

I. CELLULAR—Without regular vessels, but composed of fibres which sometimes cross and interlace each other. The *Confervæ* (green scum-like aquatic growths), the *Lichens* (which incrust stones and decaying trees), the *Fungi* (or mushroom tribe), and the *Algæ* (or sea-weeds), belong to this division. In some of these families there are no apparent seed-organs. From their mode of growth—viz., sprout-like increase of the same organ—they are known as THALLOGENS or AMPHIGENS.

II. VASCULAR—With vessels which form organs of nutrition and reproduction. According to the arrangement of these organs, vascular plants have been grouped into two great divisions—CRYPTOGAMIC (no visible seed-organs), and PHANEROGAMIC (apparent flowers or seed-organs). These have been further subdivided into the following classes :—

    1. CRYPTOGAMS—Without perfect flowers, and with no visible seed-organs. To this class belong the *mosses, equisetums, ferns,* and *lycopodiums*. It embraces many fossil forms allied to these families. From their mode of growth—viz., increase at the top or growing point only—they are known as ACROGENS.

    2. PHANEROGAMIC MONOCOTYLEDONS—Flowering plants with one cotyledon or seed-lobe. This class comprises the *water-lilies, lilies, aloes, rushes, grasses, canes,* and *palms*. In allusion to their growth, by increase within, they are termed ENDOGENS.

    3. PHANEROGAMIC GYMNOSPERMS—This class, as the name indicates, is furnished with flowers, but has naked seeds. It embraces the *cycadeæ* or pine-apple tribe, and the *coniferæ* or firs. In allusion to their naked seeds, these plants are also known as GYMNOGENS.

    4. PHANEROGAMIC DICOTYLEDONS — Flowering plants with two cotyledons or seed-lobes. This class embraces all forest trees and shrubs—the *compositæ, leguminosæ, umbelliferæ, cruciferæ*, and other similar orders. None of the other families of plants have the true woody structure, except the *coniferæ* or firs, which seem to hold an intermediate place between monocotyledons and dicotyledons; but the wood of these is readily distinguished from true dicotyledonous wood. From their mode of growth—increase by external rings or layers—the dicotyledons are termed EXOGENS.

NOTE EXPLANATORY.

SPERMOCARPS (Gr. *sperma*, seed, and *karpos*, fruit).—Literally, "fruit-seeded;" plants whose seeds contain an embryo, in which the rudiments of the future plant are distinguishable.

SPOROCARPS (Gr. *spora*, a germ, and *karpos*).—Literally, "produced by germs;" plants which have no seed-fruits, but which are reproduced by a development of certain germs or parts of their cellular tissues, called *spores*.

PHANEROGAMS (Gr. *phaneros*, apparent, and *gamia*, marriage).—Plants having apparent flowers or seed-organs.

CRYPTOGAMS (Gr. *kryptos*, concealed, and *gamia*).—Plants having no apparent seed-organs, or whose organs of reproduction are not essentially different from the other parts.

ANGIOSPERMS (Gr. *angeion*, a vessel, and *sperma*, seed).—Plants having their ovules contained in ovaries.

GYMNOSPERMS (Gr. *gymnos*, naked, and *sperma*).—Plants having their ovules in open carpels; literally, "naked or unenclosed seeds."

ANGIOSPORES (Gr. *angeion* and *spora*).—Plants having spores formed in cases which are not open till ripe.

GYMNOSPORES (Gr. *gymnos* and *spora*).—Plants having their spores superficial, and not enclosed in cases.

EXOGENS (Gr. *ex*, out, and *gennao*, I produce).—Plants whose stems increase by external layers of annual growth, as the beech and oak.

ENDOGEN (Gr. *endon*, within, and *gennao*).—Plants whose stems increase from within, by a coalescence of the footstalks of the leaves, which always encircle the growing point, as the palms and canes.

ACROGEN (Gr. *akros*, the summit).—Plants which increase by growth of the top or growing point, as the ferns, &c.

AMPHIGENS (Gr. *amphi*, around).—Plants which increase by the growth or development of their cellular tissue on all sides, as the lichens.

DICOTYLEDONS (Gr. *dis*, two, *cotyledon*, seed-lobe).—Plants whose seeds have two lobes, as the bean.

MONOCOTYLEDON (Gr. *monos*, one, and *cotyledon*).—Plants whose seeds have only one lobe, as the grasses.

THALLOGENS (Gr. *thallos*, a sprout).—Plants whose spores are attached to the frond or leaf, as the ferns.

AXOGAMS (Gr.)—Plants having their spores on a stem or axis, as the mosses and liverworts.

HYDROPHYTES (Gr. *hydor*, water, *phyton*, a shoot).—Water-plants, like the sea-weeds and confervæ.

AEROPHYTES (Gr. *aër*, the air).—Growing in the air, as the lichens, in contradistinction to the hydrophytes.

HYSTEROPHYTES (Gr. *hysteros*, the last).—The lowest or last of the plant-race, as the fungi or mushrooms.

Throwing these various groups into diagrammatic form, we have first the Amphigens—the fungi, lichens, and sea-weeds—whose homogeneous structure and simple modes of

Amphigenous Aspect of Vegetation.

growth are readily recognisable, even by the unscientific observer. Lowly alike in their aspect and functions, they cluster, as fungus-growths, over the decomposition and decay of organised tissues; mantle, as lichens, the surface of the weathering rock and the mouldering trunk; clothe, as sea-weeds, the shelves and ledges of the shallower ocean, or spread scum-like over the surface of the stagnant pool. Decay and putrescence seem to be their appointed elements: and wherever the organic cell is on the verge of dissolution into inorganic matter, there they are ready to appropriate and reconvert it once more into the circle of vitality. The pioneers of the higher orders, they elaborate a soil for their growth; cosmopolitan in habit, they are found where other plants are unknown. Such are the Amphigens now; does the palæontologist exceed his warrant when he presumes that such they ever have been from the moment they first

clustered over the rocks or spread their leathery lobes in

the waters? Next in order come the Acrogens—the mosses, equisetums, and ferns—the lovers of the swamp and shade, and the colonists of emerging and new-formed lands. Of rapid and widespread growth, they have ever contributed to the consolidation of alluvial soils, and their remains mingle largely with the coals and shales of the past, as they

Gymnogenous Aspect of Vegetation.

do now with the peat-bogs and mud-silts of the present day. Less cosmopolitan than the amphigens, they still have an extensive range ; but, like them, their function is largely physical, and comparatively few of the animal races find subsistence on their stems or foliage. As the peaty marsh, the silty lake, and the shady river-swamp are now their established headquarters, so the increment and consolidation of these by their annual growth and decay has ever been their geological function. Higher than these, and of more varied aspect, come the Gymnogens—the cycads, and yews, and pines—the gregarious forest growths of the present, as of former ages. Lovers of the temperate and coldly temperate zones—inhabitants alike of the swamp, the arid plain, and the mountain—they exhibit an enlarged diversity of habit, and form, and function. Like the acrogens, many of them are swamp and coal formers; and, as will be afterwards seen, it is to the acrogens and gymnogens, and especially to extinct intermediate forms, that we are chiefly indebted for the coal-beds of the earlier formations. As foodsuppliers, their function is comparatively limited—their dry rigid foliage, their scaly seeds and fleshless berries, being little fitted for the miscellaneous requirements of the higher animals. And it is a curious coincidence that so few of the higher animals appear in the geological periods where these acrogenous and gymnogenous groups so universally prevail. The Endogens—the grasses, lilies, and palms—follow next in order, and present a still increasing variety, both in form, habitat, and function. Tropical and temperate, but unfitted for the extremes of climate, they assume more diversified areas of localisation, and become more and more fitted for the sustenance of a varied terrestrial fauna. While radiates, molluscs, and crustacea may feed on the thallogens, and insects, and it may be a few birds and reptiles, find their food and shelter among the acrogens and gymnogens, it is certainly to the endogens and exogens that the higher terrestrial animals turn for their main depend-ence. The formative or geological function so prominent in the lower groups, now gives place to the alimentative;

Endogenous Aspect of Vegetation.

and though the grassy carpet may conserve the soil from waste, and the palm-grove may induce the accumulation of vegetable matter, still the relations of the endogens are mainly and obviously zoological. Highest and last come

the Exogens—the herbs, and shrubs, and timber-trees—
which, in their beauty and variety and dignity of aspect,

Exogenous Aspect of Vegetation.

crown the long line of vegetable existences. Slower of
growth, but of greater longevity, the beauty of their flowers,
the utility of their seeds and fruits, the durability of their
structure, and the diversity of their habits and forms, all
point to them as the culminating orders of the vegetable

kingdom. And it is curious to learn that, unknown in the earlier eras, and just beginning to make their appearance in the secondary epochs, they come into full force and vigour in the tertiary and post-tertiary—the periods at which the higher animals and man are present to reap the advantages of their more varied utilities.

Such are the leading features of the great groups of the vegetable kingdom—groups to which we shall have frequent occasion to allude when we come to treat the successive stages of the fossil flora, and which are here displayed in pictorial outline with a view to facilitate the comprehension of these allusions. Though thus arranged in physiological groups, the whole, from the simple cell that floats on the putrid pool to the noblest tree of the forest, forms but one orderly and co-adjusted system; and could we combine the extinct with the living, the same order and co-adjustments would be found to run as unswervingly through the wider combination. The conception is one, though its expression through time and space must necessarily assume the character of infinite diversity.

Subdividing still further, according to their most marked characteristics, whether external or internal, the botanist arranges all the forms of vegetable life into some 60 or 70 orders, about 300 genera, and upwards of 100,000 species. As most of these distinctions, however, are founded on the form and connection of the flower, fruit, and leaf—organs which rarely or never occur in intelligible union and preservation in a fossil state—the palæontologist is guided in the main by the great structural distinctions already adverted to, and not unfrequently by the simple but unsatisfactory test of "general resemblance." On the whole, Fossil Botany, or Palæophytology, as it is sometimes termed, is by no means in a satisfactory state, and the science languishes for the advent of some master minds to do for it what Cuvier and Agassiz and Owen have done for the sister science of Fossil Zoology.

Notwithstanding the fragmentary state of the plants that turn up to the geologist, the greatly altered conditions of the parts that are found, and the hopelessness of ever discovering the legible dispositions of such evanescent portions as the floral organs, on which so much of existing botany is founded: notwithstanding all these obstructions, there is still so much remaining—the structure of the roots, stems, barks, leaves, fronds, and fruits—the characteristic markings of their different surfaces—and the scars which their parts leave on separation—that the competent botanist, armed with his microscope and ample means of comparison, should have little difficulty in arriving at many definite and important conclusions. The anastomosing disposition of a sea-weed is surely sufficiently distinct from the branching aspect of a terrestrial plant—the reticulate venation of a dicotyledonous leaf from the parallel arrangement of a monocotyledon—the scalariform tissue of a fern from the punctated tissue of a conifer—and the bundled mass of an endogenous stem from the concentric layers of an exogen. These and many other characteristics are sufficiently preserved in the strata of every formation; and though we may not be enabled to say, on the principles of existing botany, that this fragment is that of a cruciferous plant, and that of a leguminous one, we have, at all events, enough to fix in the mean time the great progressional order of plant-life from the predominance of *Acrogenous* orders in primary formations to the higher *Gymnosperms* of the secondary, and from these again to the still higher *Angiosperms* of the tertiary and current epochs. And Geology, strong in the faith of Nature's unity and persistency of plan, rests assured, that under right methods of research the key to that Plan will yet be discovered, enabling the palæontologist to unfold the relations of fossil plant-life, its distribution in space, and its progress in time, even as the botanist now determines its existing relationships, and maps out its centres and areas of geographical arrangement.

2.—ITS FAUNA OR ANIMAL LIFE.

As with plants, so with animals. While we find them everywhere—on the earth, in the air, and in the waters—on the substances of plants, and even in the living tissues of other animals—they are as imperatively governed by the influences of climate, food, and other external conditions as the Vegetable world, though possessed for the most part of a locomotion which at first sight might seem to confer on them an ubiquity of habitat. Thus, the FAUNA of the tropics is essentially different from that of the temperate zone, and the animals which people the temperate zone have but little in common with those of the arctic regions. It is true that some, like Man and his companions, the dog, horse, and other domesticated animals, have a range all but universal; but generally speaking, the zones of Animal Life—horizontally and vertically—are about as sharply defined as those of vegetation. The elephant and rhinoceros that luxuriate in the low tropical jungle would fare but indifferently on the lofty slopes of the Himalayas; while the buffalo and bison which herd at these heights would cease to exist were they raised but a few thousand feet higher. As with *altitude* on land, so with *depth* in the ocean; and thus the sea-weeds and shells that grow and live within the influence of the tides constitute a *Littoral* zone very different from the *Laminarian* or broad sea-tangle zone which extends, in British seas, from 40 to 90 feet in depth; this again is essentially distinct from the *Coralline* zone, which ranges from 90 to 300 feet, and is the great theatre of marine life; while beyond this lies the *Coral* zone, the region of the strong calcareous corals extending from 300 to 600 feet in depth from the shore line. But it is not alone to climate and external conditions that we must look for the variety and distribution of animal life. There is an aboriginal diffusion of different tribes and families from certain centres and over certain areas, for which science can as yet offer no satisfactory reason. Thus, why should the giraffe, or ostrich, or hippopotamus, be restricted to the continent of Africa, while the forests, and plains, and river-swamps of South America enjoy the same tropical sun, and seem every way equally adapted to identity of vitality? The pampas of America, as has been proved by experience, are as well fitted for the increase of the horse as the plains of Europe or the steppes of Tartary; and yet, till man carried him thither a few hundred years ago, no horse of the current epoch existed there. The ornithorhynchus burrows only in the river banks of Australia; the apteryx is unknown beyond the limits of New Zealand; the sloth is confined to the tropical forests of America; the armadillo to the same region; and not one of the Old World monkeys is identical with any of those of the New. Nor is it alone the terrestrial tribes that are thus limited and restricted; the aërial and aquatic, though possessing superior facilities for dispersion, are equally circumscribed, each within its own geographical habitat. The humming-birds flutter only over the flowers of the New World; the pheasants are unknown beyond the coverts of the Old; the shark-like cestraciont frequents alone the waters of the Southern Pacific; and the trigonia never carries its shell beyond the shores of Australasia. Such restrictions we cannot explain unless by ascribing them to independent centres of creation, or to means of distribution that prevailed during former geological epochs, but which ceased to exist when sea and land received their present relations. And this brings us to remark on what are termed by zoologists the law of identity and the law of representation; that is, that different regions, though not peopled by identical species, may be peopled by animals which perform analogous functions, and represent them, as it were, in the great plan of vital economy. Thus, the ostrich of Africa is represented in South America by its congener the rhea; the jaguar and puma of the New World represent the tiger and lion of the Old; the camel of Arabia finds its analogue in the llama of Peru; and similar functions are at once discharged by the gavial of the Ganges, the crocodile of the Nile, and the alligator of the Amazon. Over and above these physical relationships there is also that which has reference to the size of the animal, and the element in which it is destined to live. As a general rule, and each within its own order or family, the aquatic members are larger than the terrestrial; the amphibious bulkier than those that are strictly terrestrial; the marine superior in size to those of fresh-water habitat; and the terrestrial more massive than the arboreal. Admitting these relations, and reasoning from the present to the past, the comparative bulk of organic remains may often become an index to external conditions of life, and throw light over the investigations of the palæontologist, when other indications are uncertain and obscure.

Besides these distinctions and restrictions imposed on vitality by external conditions, there are those connected with the functions they have to perform in the economy of nature. Some, for instance, are fitted to live on a purely vegetable diet, others to prey on the flesh of other creatures; some are constructed so as to feed only on seeds and grains, others to prey solely on insects; many earn their subsistence by a life of ceaseless activity and toil, others are formed for parasitic attachment to the living tissues of larger animals, and there find life and enjoyment without a single effort or care of their own. And as these varied functions necessarily require for their performance a special adaptation of organs—a tooth to cut or a tooth to grind, a foot to seize or a foot to dig, a limb to run or a limb to fly—so will similar modifications afford to the palæontologist an evidence of functions performed in bygone ages, and enable him, not only to reconstruct forms of harmonious organs, but to assign to these organs the part they had to play in the great drama of vitality. In the performance of these varied functions many animals have to make long periodic migrations, either for the immediate purpose of procuring food and shelter for themselves, or prospectively for their future young. From colder to warmer regions, and from warmer to colder—from land to water, and from water to land—from sea to river, and from river to sea—there is ever, among certain animals, an incessant interchange; and though palæontology has yet been unable to detect such migrations in the past, we may rely on their occurrence, and be prepared to admit the fact into our inferences and reasonings.

Coexistent with and beyond all this, there are those innumerable differences of species and kind and family and class, which we can only resolve into the eternal will of the Creator. Why, for instance, should the polype differ from the star-fish, the star-fish from the crab, the crab from the turtle, the turtle from the fish, the fish from the bird, or the bird from the quadruped? It is in vain to tell us that the one is but a progressive or developmental form of the other—that the reptile is but a transmutation, in time and under new external conditions, from the fish, and that the fish is but the lineal descendant of the shell-fish. Admitting that such was the true genetic origin of the various grades of vitality, there still lies behind and unaccounted for the orderly plan in which such development shall occur, and the reason for the definite specific forms which the descendants invariably assume. Grant, we again repeat, that all vitality were indissolubly interwoven into one great genetic mesh, still that mesh presents, at determinate times and over determinate areas, definite variety and speciality of pattern. Whence this orderly variety? Wherefore these special and distinctive patterns? At the most, Science can only note the distinctions, it can never hope to assign the reason. To do so would be to place the intelligence of the finite creature on the same level with the prescience of the infinite Creator. It is our high privilege, however, to observe and reason; and, reasoning, to arrange and classify the animal kingdom according to their different grades and affinities, and so arrive at some intelligible comprehension of the great scheme of vitality.

As in Botany, so in Zoology this arrangement is greatly facilitated by the fact that, numerous as animal forms are, they are all constructed after a few primal types and patterns. Some are furnished with a bony skeleton, the leading feature of which is the vertebral column or backbone—these are the VERTEBRATES; others have no such osseous framework—these constitute the INVERTEBRATES. As the *leaf* was the primary organ in the plant's development, so the *vertebra* seems to be the primal organ in the vertebrate skeleton; and by its modifications and adaptations for special ends, the Creator has produced every form of terrestrial, aërial, and aquatic existence. According to the modern doctrines of anatomy, the skull, or brain-case, is composed of vertebral bones, modified and adapted for a special purpose: so are the limbs, whether for running, flying, or swimming; so also the ribs, whatever their form or number; and in like manner all the other appurtenances of the vertebrate skeleton. This is the great doctrine of Homology, or science of similar parts, as it is termed, through which we arrive at the conclusion that the arm and hand of man, the fore-limb and foot of the quadruped, the wing of the bird, and the fore-fin of the fish, are one and the same primal organ, composed of the same or *homologous* parts, and merely modified or altered for the performance of certain special functions. As the stationary engine that turns the spindles of the factory, the locomotive that drags the railway cars, and the marine engine that propels the steamship, are but modifications of the same primal machine; so the mammal that runs, the mammal that flies, and the mammal that swims, are but specialised expressions of the same primal plan, the creation of a new type being unnecessary where a modification of an existing one would suffice. Knowing these modifications in the limbs, jaws, teeth, and other organs, and the ends they were meant to subserve in living races, we can predicate of forms long since extinct, and can associate with co-relation of structure the functions that creatures were meant to perform in the economy of former ages. It is by this "law of the co-relation of parts," and faith in the uniformity of nature's method, that Cuvier and Owen, and other great anatomists, have been enabled to accomplish their wonderful restorations of extinct life, and from a few sorely mutilated and scattered fragments to present us with forms of harmonious entirety. "Every organised being," says the great French anatomist, "forms a whole, a single circumscribed system, the parts of which mutually correspond and concur to the same definite action by a reciprocal reaction. None of these parts can change without the others also changing, and consequently each part, taken separately, indicates and gives all the others." As with the vertebrate type, so with the molluscan, the articulate, and the radiate. There is a plan and primal pattern to each, and that plan, modified and specialised, can be traced through every species and individual of the division, no matter how varied and numerous they may be. And what has been done to homologise the external framework will shortly be done for the muscular, respiratory, and vascular systems—for the organs of digestion, secretion, and reproduction—so that we may no longer combine things that are merely *analogous* with those that are *homologous*, and thus confound, in our interpretations of nature, beings that were from the first constructed on an essentially different basis.

Proceeding on grounds such as these, the zoologist separates the vertebrate from the invertebrate, the mammals from the birds, the birds from the reptiles, and the reptiles from the fishes. He also separates the invertebrate shell-fish from the invertebrate crab, the crabs from the sea-urchins, the sea-urchins from the star-fishes, the star-fishes from the corals, and these again from the lower sponges that can scarcely be distinguished from the sea-weeds that surround them. Looking at the manner in which the functions of *nutrition, reproduction,* and *sensation* are performed in each of these classes, we speak of "higher" and "lower" forms, of creatures of more simple and of more complex organisation; but we do not say—and reason and experience alike shrink from endorsing the allegation—that one form or family is less perfect than another, either in its nature or in the functions it was designed to perform.

> "All are but parts of one stupendous whole,
> Whose body Nature is, and God the soul;
> That changed in all, and yet in all the same;
> Great in the earth as in the ethereal frame;
> Warms in the sun, refreshes in the breeze,
> Glows in the stars, and blossoms in the trees;

> Lives through all life, extends through all extent,
> Spreads undivided, operates unspent;
> Breathes in our soul, informs our mortal part,
> As full, as perfect in a hair as heart;
> As full, as perfect in vile man that mourns,
> As the rapt seraph that adores and burns;
> To him no high, no low, no great, no small,
> He fills, he bounds, connects, and equals all."

While thus disclaiming the idea of "imperfection" as applicable to any grade of vitality, it would be erring against all reason and instinct to discard the terms "higher" and "lower" in treating of organised existences. The creature consisting of a uniform mass must appear, even to the most untutored observer, to stand on a humbler platform than that composed of a variety of parts and tissues. The protozoan, that envelops its food in its gelatinous sac, assimilates the nutritive juices, and then rejects the remainder, and this without mouth, stomach, or opening of any kind, is certainly *lower* (*or less highly organised*, if you will) than the mollusc, which is furnished with mouth, stomach, and alimentary canal; and the mollusc, furnished only with external gill-tufts and the merest heart-like cavity, can never be placed on the same level with the quadruped provided with masticating and salivatory apparatus, its stomach, its organs of chylification and chymification and intestinal canal—its respiratory and circulating system of lungs, heart, veins, and arteries. Again, the protozoan that reproduces itself by a mere cellular expansion of its own mass—a mass, any portion of which is equally vital, and capable of becoming a separate creature—is surely lower in the scale than the shell-fish that reproduces by spawn, and would perish under subdivision of its tissues; while the reptile, reproducing by eggs, which it drops in the stagnant pool and never cherishes, can never, without the abuse of everything like discrimination, be ranked so high as the mammal that brings forth its young alive, and even then, by a special organisation, suckles them during months with assiduous care. But on such distinctions we need not dwell. They were made long before observation had shaped itself into systems of science, and are patent alike to the learned and the unlearned. This dictum, therefore, the zoologist lays down, *that the lower a creature is in the scale of being, the more its individual parts resemble each other* (vegetative repetition); *and the higher it is, when, instead of several functions being performed by the same organ, each function, be it of nutrition, reproduction, or sensation, is performed by an organ specially devoted to it.*

This brings us to the classification of the zoologist; and in comparing the Past with the Present Life of the Globe, the palæontologist requires to invent no new system or scheme of arrangement. One plan and design runs through the whole of animated nature; and though species and genera, and even whole families, have died out, and others have taken their places—and this has been repeated again and again—still have all the successive incomers been constructed upon the same plan, and designed to perform analogous functions. The classification of the palæontologist is therefore the same as that of the zoologist, with the exception of such extinctions as fill up the gaps that exist between conterminous genera, and render more compact and harmonious, if we may so speak, the grand scheme of terrestrial vitality. The following outline of the animal kingdom will render more intelligible the comparisons we have to institute between the past and the present—between the forms that now live and act, and those that have become extinct and been converted into stone thousands of ages ago :—

VERTEBRATA,

Or animals with back-bone and bony skeleton, and comprehending
MAMMALIA, AVES, REPTILIA, and PISCES.

I. MAMMALIA, or *Sucklers*, subdivided into Placental and Aplacental.

1. PLACENTAL, bringing forth mature young.

BIMANA (*Two-handed*)—Man.
QUADRUMANA (*Four-handed*)—Monkeys, Apes, Lemurs.
CHEIROPTERA (*Hand-winged*)—Bats, Vampyre-bats, Fox-bats.
INSECTIVORA (*Insect-eaters*)—Mole, Shrew, Hedgehog, Banxring.
CARNIVORA (*Flesh-eaters*)—Dog, Wolf, Tiger, Lion, Badger, Bear.
PINNIPEDIA (*Fin-footed*)- Seals, Walrus.
RODENTIA (*Gnawers*)—Hare, Beaver, Rat, Squirrel, Porcupine.
EDENTATA (*Toothless*)—Ant-eater, Armadillo, Pangolin, Sloth.
RUMINANTIA (*Cud-chewers*)—Camel, Llama, Deer, Goat, Sheep, Ox.
SOLIDUNGULA (*Solid-hoofs*)—Horse, Ass, Zebra, Quagga.
PACHYDERMATA (*Thick-skins*)—Elephant, Hippopotamus, Rhinoceros.
CETACEA (*Whales*)—Whale, Porpoise, Dolphin, Lamantin.

2. APLACENTAL, bringing forth immature young.

MARSUPIALIA (*Pouched*)—Kangaroo, Opossum, Pouched Wolf, &c.
MONOTREMATA (*One-vented*)—Ornithorhynchus, Porcupine-ant-eaters.

II. AVES, or BIRDS.

RAPTORES (*Seizers*)—Eagles, Falcons, Hawks, Owls, Vultures.
INSESSORES (*Perchers*)—Jays, Crows, Finches, Sparrows, Thrushes, &c.
SCANSORES (*Climbers*)—Woodpeckers, Parrots, Cockatoos, &c.
COLUMBÆ (*Pigeons*)—Common Dove, Turtle Dove, Ground Dove.
RASORES (*Scrapers*)—Barnfowl, Partridge, Grouse, Pheasant.
CURSORES (*Runners*)—Ostrich, Emeu, Apteryx.
GRALLATORES (*Waders*)—Rails, Storks, Cranes, Herons.
NATATORES (*Swimmers*)—Divers, Gulls, Ducks, &c.

III. REPTILIA, subdivided into Reptiles Proper and Batrachians.

1. REPTILES PROPER.

CHELONIA (*Tortoises*)—Turtles, Tortoises.
LORICATA (*Covered with Scales*)—Crocodile, Gavial, Alligator.
SAURIA (*Lizards*)—Lizard, Iguana, Chameleon.
OPHIDIA (*Serpents*)—Vipers, Snakes, Boas, &c.

2. BATRACHIANS, or FROGS.

ANOURA (*Tail-less*)—Toad, Frog, Tree-frog.
URODELA (*Tailed*)—Siren, Triton, Salamander.
APODA (*Footless*)—Lepidosiren, Blindworm.

IV. PISCES, or FISHES.

SELACHIA (*Cartilaginous*)—Chimæra, Sharks, Sawfish, Rays.
GANOIDEA (*Enamel-scales*)—Amia, Bony-pike, Sturgeon.
TELEOSTIA (*Perfect-bones*)—Eels, Salmon, Herring, Cod, Pike, &c.
CYCLOSTOMATA (*Circle-mouths*)—Lamprey.
LEPTOCARDIA (*Slender-hearts*)—Amphioxus.

INVERTEBRATA,

Or animals void of back-bone and bony skeleton, and comprehending

ARTICULATA, MOLLUSCA, RADIATA, and PROTOZOA.

I. ARTICULATA, subdivided into Articulates and Vermes.

1. ARTICULATA, or Jointed Animals Proper.

INSECTA (*Insects*)—Beetles, Butterflies, Flies, Bees.
MYRIAPODA (*Many-feet*)—Scolopendra, Centipedes.
ARACHNIDA (*Spiders*)—Spiders, Scorpions, Mites.
CRUSTACEA (*Crust-clad*)—Crayfish, Crabs, Shrimps, Woodlice.
CIRRHOPODA (*Curl-feet*)—Acorn-shells, Barnacles.

2. VERMES, or Worms Proper.

ANNELIDA (*Small-rings*)—Lobworm, and almost all the marine worms.
ROTIFERA (*Wheel-bearers*)—Rotifers, Hydatina.
GEPHYRIA (*Intermediates—urchin-like*)—Sipunculus, Echinurus.
LUMBRICINA (*Earth-worms*)—Earth-worms, Nais.
HIRUDINEI (*Leeches*)—Leeches, Branchellion.
TURBELLARIA (*Turbellaries*)—Planaria, Ribbon-worms.
HELMINTHES (*Gut-worms*)—Intestinal worms.

II. MOLLUSCA, subdivided into Mollusca and Molluscoida.

1. MOLLUSCA, or Shell-fish Proper.

CEPHALOPODA (*Head-footed*)—Cuttle-fish, Octopus, Calamary, Nautilus.
PTEROPODA (*Wing-footed*)—Clio, Hyalæa.
GASTEROPODA (*Belly-footed*)—Snails, Slugs, Whelks, Cowries.
ACEPHALA (*Headless*)—Oysters, Mussels, Cockles, Shipworms.
BRACHIOPODA (*Arm-footed*)—Terebratula, Lingula.

2. MOLLUSCOIDA, or Mollusc-like Animals.

TUNICATA (*Coated, but Shell-less*)— { Biphora, Simple and Compound Ascidians.
POLYZOA (*Compound animals*)
  or
BRYOZOA (*Moss-like animals*) } Flustra, Eschara, Plumatella. &c.

ITS FAUNA. 59

III. RADIATA, or ZOOPHYTES—Ray-like Animals.
ECHINODERMATA (*Urchin-skinned*)—Sea-urchins, Star-fishes.
ACALEPHÆ (*Sea-nettles*)—Jelly-fish, Beroës.
POLYPI (*Many-feet*)—Coral animals, Sea-anemones, Hydras.

IV. PROTOZOA, or LOWEST-LIFE—Globular Animals.
INFUSORIA (*Infusories*)—Monads, Volvoces, Vorticella.
PORIFERA (*Pore-bearers*)—Sponges, Fresh-water Sponges.
RHIZOPODA (*Root-footed*)—Amœba, Polythalamia (Foraminiferæ).

Throwing, as in the case of the vegetable world, these great groups into diagrammatic form, we have first the

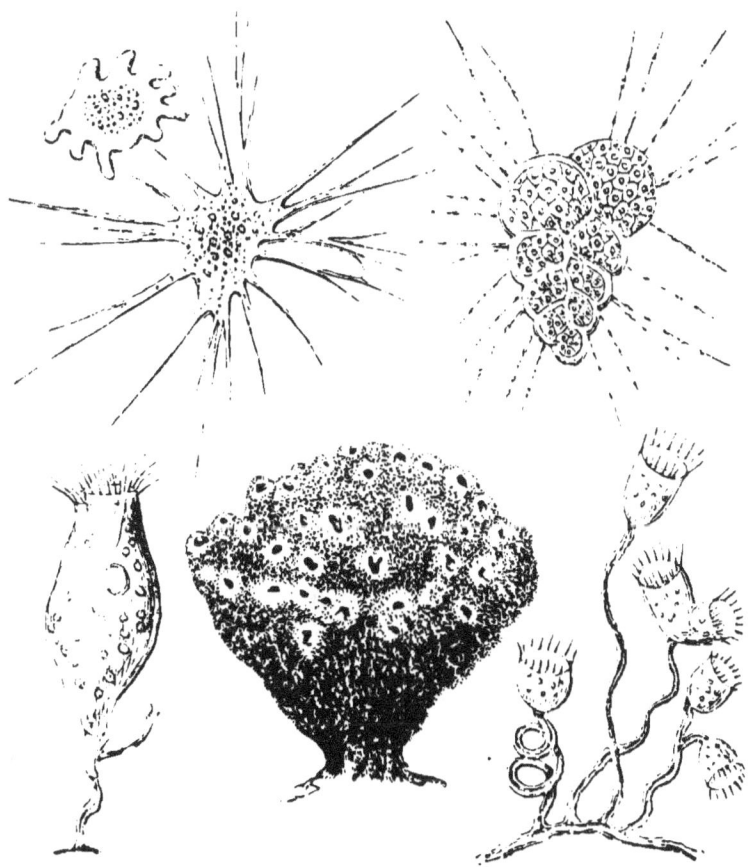

Protozoan Aspect of Animal Life.

Protozoa—the sponges, foraminiferæ, and infusorial animalcules—which, half-plant half-animal, stand, as it were, on the verge of organised existence. Restricted to the waters, rooted as sponges to the sea-bed, appearing as infusoria (we cannot tell how) in stagnant and putrid waters, or thronging in inconceivable numbers as foraminiferæ alike the shallow estuary and the profoundest ocean-depth, their office seems to be the reconversion of organic matter from ultimate decay, and the reconstruction of mineral matter from a state of solution and diffusion. Mere gelatinous specks or glairy films, encased in or encasing some horny, flinty, or limy framework, they constitute the food of many of the lower orders, though their function, on the whole, is

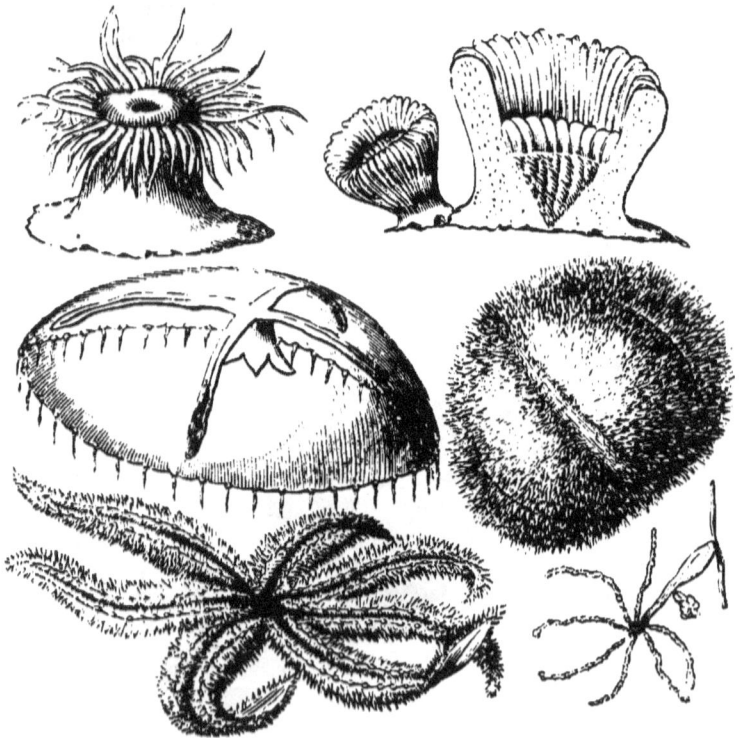

Radiate Aspect of Animal Life.

mainly formative or geological. As the calcareous muds of

existing seas and estuaries are in great part composed of the shelly coverings of the minute foraminiferæ, and the siliceous muds composed of the still minuter shields of infusoria, so we shall afterwards find that extensive strata in the earth's crust owe their formation to similar agencies. Next above these lowest forms of life stand the Radiata— the corals, sea-anemones, jelly-fish, star-fishes, and urchins —all, too, inhabitants of the ocean, and in one or other of their orders appearing in every depth and in every latitude of space. Elaborating from microscopic organisms, the material of their pulpy fabrics, which in turn become the food of the higher orders, their function, though more largely biological than the protozoans, is still in a great measure formative. To coral-zoophytes we owe our existing coral reefs, and from the same source, or from their allies the encrinites, have sprung many of the massive limestones that give character to the crust of the globe. The office of the radiata is thus comparatively humble, as their organisation, though beautifully symmetrical, is simple and lowly. Next we approach the Molluscoida, or mollusc-like organisms of modern naturalists—the sea-mats and dead-men's fingers of the fisherman and common observer. Fixed in their habitats, and elaborating, like the corals and sponges, their structures from the waters of the ocean, their functions are humble and their characters obscure. From them we ascend in the zoological scale to the true Mollusca—the oysters, mussels, cockles, whelks, snails, slugs, nautili, and cuttle-fishes—the "shell-fish" of everyday language, though many of them are naked and altogether shell-less. Of more diversified organisation than any of the preceding groups, they are, in one or other of their orders, inhabitants of the ocean, the lake, the river, the marsh, and the dry land. Having also a more cosmopolitan range—feeding, some on plants and others on animals—being in turn preyed upon by other races, aquatic, aërial, and terrestrial, and creating extensive calcareous masses with their shelly coverings—the mollusca fulfil im-

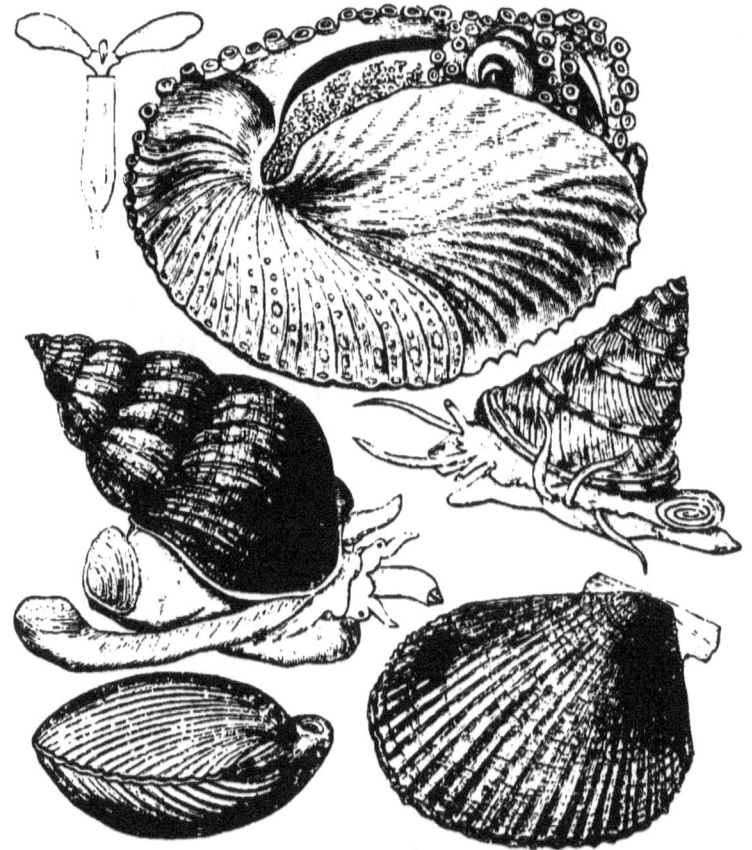

Molluscan Aspect of Animal Life.

portant vital as well as physical functions. From the enduring nature of their testaceous coverings, they become important indices to the palæontologist, and the interpretations of geology have largely profited by the persistency of their remains. Next in order come the Articulata—the worms, insects, spiders, and crabs—known at once by their many-ringed, segmented, or jointed bodies. Inhabitants of every element—earth, air, and ocean—and even finding their abodes as parasites on other animals, the articulata have a function as diversified as their organisation. They

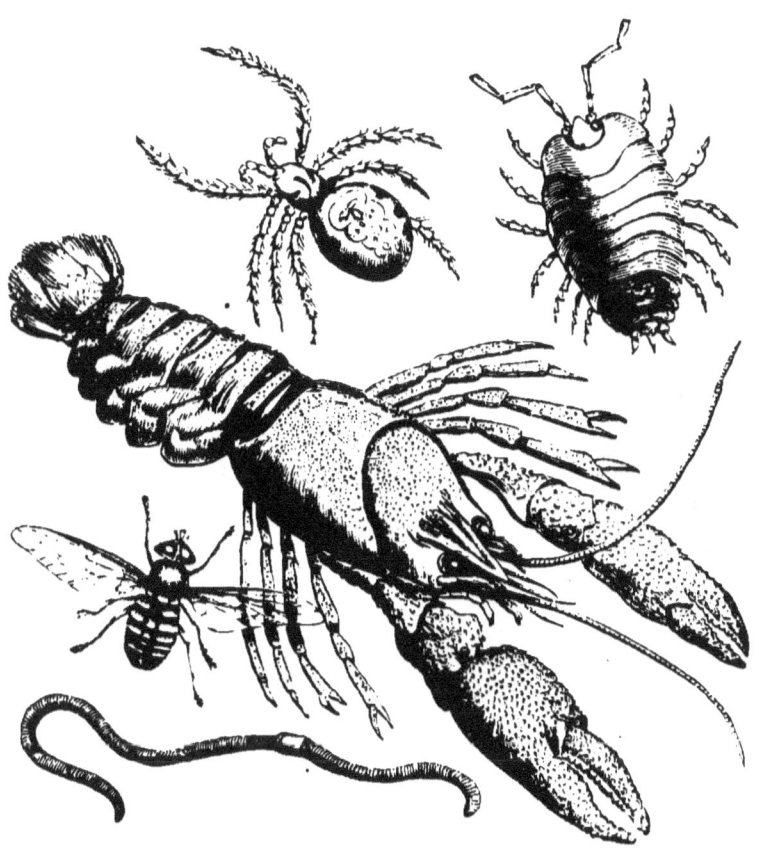

Articulate Aspect of Animal Life

are vegetable as well as animal feeders, and occur in every region, though culminating in numbers, size, and specific variety under the genial influences of equatorial and tropical latitudes. Their world-office is mainly biological; and while preying alike on plants and animals, they become in turn the principal food of other creatures—fishes, birds, reptiles, and mammals. Lastly, and highest and most diversified in structure, come the Vertebrata—the fishes, reptiles, birds, and mammals—the inhabitants of every element, and the tenants of every region, though culminating chiefly in numbers and rank within temperate and warm parallels.

Vertebrate Aspect of Animal Life.

In general terms, the ascent in the zoological scale is from the aquatic to the terrestrial, from the cold-blooded water-breathers to the cold-blooded air-breathers, and from these to the still higher warm-blooded air-breathers. The great majority of invertebrate forms are confined to the waters : a large proportion of the vertebrates are strictly terrestrial, or own an amphibious existence. Between them and the higher forms of terrestrial vegetation, the interdependence is complete ; the existence of the higher flora being unintelligible in the absence of a higher fauna. The grasses, foliage, seeds, fruits, and roots of the one kingdom become the indispensable sustenance of the vegetable-feeders of the other; while the vegetable-feeders in turn become the food of the carnivora. Among the vertebrata the actions and reactions of life are more immediate and apparent, and in them alone are manifested all the higher offices of vitality. Sense, instinct, volition, reason, and moral perception, mark the line of ascent. The vital predominates over the material, and in the culminating order (Bimana) the psychological rise superior to the physiological functions.

Such are the leading divisions of the animal kingdom, which are again divided and subdivided into families, and genera, and species—each minor group presenting a distinct and determinate pattern on the great web of created existence. By a study of these patterns, and a knowledge of their manifold relations, the zoologist is enabled to arrive at some intelligible idea of the scheme of existing vitality; and so, possessed of similar knowledge, the palæontologist strives to reunite his scattered fragments, and to assign to them their proximate place in the still greater scheme which combines the present with the past, and the forms that have become extinct with those that still flourish around us. In the study of Fossil Zoology, or Palæo-Zoology, as it is termed, much more satisfactory progress has been made than in the sister department of Fossil Botany, the harder structures of animals (corals, shells, crusts, scales, scutes, teeth, and bones) being better preserved than the softer and more perishable tissues of vegetation. It is true that many of these fragments are widely scattered and sorely mutilated, that marine forms are relatively more abundantly retained than those of terrestrial origin, that only the merest specks of the fossiliferous strata have yet been examined, and that the sea now rolls over stratified areas vastly more extended than those that lie patent to geological research. Still, in face of all these obstructions and imperfections, palæontology has wonderfully enlarged our conceptions of vitality, has opened up to the present age a theme altogether unknown to our ancestors, and, guided by a true knowledge of the present, is destined yet to unfold a fuller and fairer vision of the life that has gone before us. As the zoologist pushes his discoveries into space, so the palæontologist pushes his discoveries into time. As the former turns to unexplored regions in the hope of finding new forms, so the latter turns to unexplored formations—formations whose areas are as varied as their dates, and whose strata give promise of other and other life-revelations for centuries yet to come.

3.—CO-ADAPTATIONS OF FLORA AND FAUNA.

Perfect as the existing flora and fauna may appear, each in its own proper line, they are only constituent portions of a greater life-system, bound together by numerous co-adaptations and adjustments. As each is adapted to, as well as dependent on, external conditions, so both are dependent on one another, and, as presently constituted, neither could possibly enjoy a separate existence. Both, for example, are incessantly dependent on the atmosphere, yet the oxygen which the plant exhales is inhaled by the animal, and the carbonic acid expired by the animal is absorbed and assimilated by the plant. The plant rooted in the soil and casting abroad its leaves and branches in the atmosphere, though seemingly deriving the main elements of its growth from inorganic sources, is nevertheless stimulated into life and exuberance by the presence of organic decay; while the animal, being herbivorous, subsists immediately upon plants, or, if carnivorous, preys upon the plant-feeders, and is thus also ultimately dependent on the vegetable world for its subsistence. The law of circulation and interdependence is complete; and no portion of the circle could be removed without a corresponding change in the characters of the vegetable and animal kingdoms. Again, many plants are dependent on the locomotive powers of animals for their wider dispersion and increase; while other animals acquire a wider range through this new and increased source of subsistence. Further, as many animals, in their habits and organisation, are altogether fitted for an arboreal existence, the destruction of the tree would involve the destruction or non-existence of this peculiar organisation; and as other creatures are specially fitted to live on certain fruits, leaves, and roots, the disappearance of these specific supplies would necessarily involve the annihilation of the consumers. As in existing nature these and many other similar adaptations are fixed and certain, and we may safely reason from cause to effect and from effect to causation, so, in the ancient world, we may rely on similar adjustments—reasoning from certain phases of plant and animal life to the conditions under which they must have existed, and from the presence of certain races of plants and animals to the existence of other plants and animals to which they were necessarily co-adapted. It is thus that the study of the Past becomes hopeful, and Palæontology assumes the character of an inductive and reliable science. The Present is ever the safest guide to the Past; the Extinct is ever most clearly illuminated by the light reflected from the Existing.

THE RECORD.

ROCK-FORMATIONS AND LIFE-PERIODS OF GEOLOGY.

IN the preceding chapter we have endeavoured to lay before the reader a brief sketch of the PRESENT LIFE OF THE GLOBE—its plants and animals; the causes which seem to affect their growth; the conditions that govern their geographical distribution; their ordinal characters, as known to the botanist and zoologist; and the functions they are apparently destined to perform in the economy of creation. We now turn to that which is extinct—to that which geology exhumes from the rocky crust, and palæontology reinvests with verdure and vitality, as it clothed the forests and peopled the fields and waters thousands of ages before the human eye was created to be gladdened by its beauties or startled by its marvels. Before we can institute a satisfactory comparison, however — before we can decide between the older and the newer, and trace the order of their incomings and their outgoings in the scheme of nature—we must first appeal to the geologist for the order, *in point of time,* that prevails among the stratified formations.

In the "crust" or accessible portion of the globe, we discover two great sets of rocks—the one massive and unstratified, like the solidified lavas of Hecla and Vesuvius, and evidently the products of *igneous* eruption; the other stratified, or occurring in layers, like the silt of lakes and seas, and undoubtedly the results of *sedimentary* or *aqueous* operations. Between these two great forces—the aqueous and igneous—the crust of the earth is ever held in habitable equipoise and never-ending variety of superficial aspect. As the former tends to waste and wear down, and carry the eroded material to the bottoms of lakes and estuaries, there to be spread out in layers of varied consistency, so the latter as incessantly strives to elevate and reconstruct—here throwing up the sea-bed into new islands, there disrupting and undulating the solid crust, and anon casting forth from volcanic craters new rocks and rocky compounds. These forces being incessantly active, such transpositions of sea and land must have frequently taken place—piling the newer deposits over those of earlier dates, varying at every turn the relative distribution of sea and land, and offering different conditions of life to plants and animals at each successive mutation. And as the sediments of existing lakes and seas envelop the remains of plants and animals that have lived in their waters, or been borne thither by floods and rivers, so also must there have been entombed in the sediments of former epochs the plants and animals of the period—the deepest being the oldest or first-formed, and the others occurring above them *in order of time* or *superposition*. This is the great key to geological sequence: the deeper, the older, and the older, the wider the difference between fossil plants and animals and those now existing. To the palæontologist this physiological difference becomes, as it were, the measure of chronological progress; stratigraphical sequence and vital gradation are but convertible terms; and either were resolvable into TIME could we only determine the ratio of its increment and advancement.

Presuming on the uniformity of nature's operations—and

without this presumption the history of the Past would be an uncertainty and delusion—the geologist proceeds to unfold the history of the stratified deposits, tracing back from the silt of yesterday's tide to the first-formed strata; and this through the lapse of ages for which chronology has no name save "cycles" and "systems" of indefinite duration. Geology is not entitled—it dare not, in the spirit of true philosophy, appeal to "abnormal conditions," to "cataclysms," or to "revolutionary forces," for a solution of its problems. Certain agents may act over certain areas with greater intensity at one period than at another, or may exert themselves, in the varying distributions of sea and land, over wider areas; still the results are homologous though differing in magnitude, and cannot be ascribed to convulsion or disorder. Where geology cannot explain, it can at least observe and describe, and this its legitimate cultivators will ever do, rather than take shelter under the assumption of abnormal conditions in primeval nature. There is ever much more philosophy in honest doubt than in the utmost ingenuity of unsupported assumption.

"The agencies," we have elsewhere* observed, "that now operate on and modify the surface of the globe; that scoop out valleys and wear down hills; that fill up lakes, and estuaries, and seas; that submerge the dry land, and elevate the sea-bottom into new islands; that rend the rocky crust, and throw up new mountain-chains; and that influence the character and distribution of plants and animals,— are the same in kind—though differing, it may be, in degree—as those that have operated in all time past. The layers of mud, and sand, and gravel, now deposited in our lakes and estuaries, and along the sea-bottom, and gradually solidifying into stone before our eyes, are the same in kind with the shales and sandstones and conglomerates that

* *Advanced Text-Book of Geology.*

compose the rocky strata of the globe; the marls of our lakes, the shell-beds of our estuaries, and the coral-reefs of existing seas, year after year increasing and hardening, belong to the same series of materials, and in process of time will be indistinguishable from the chalks, and limestones, and marbles we quarry; the peat mosses and jungle growth, and the vegetable drift that have grown and collected within the history of man, are but continuations of the same formative power that gave rise to the lignites and coals of the miner; the molten lavas of Ætna and Vesuvius, and the cinders and ashes of Hecla, are but repetitions of the same materials which now compose the basalts and greenstones and trap-tuffs of the hills around us; the corals, and shells, and fishes, the fragments of plants, and the skeletons of quadrupeds, now imbedded in the mud of our lakes and estuaries and seas, will one day or other be converted into stone, and tell as marvellous a tale as the fossils we now exhume with such interest and admiration." Without this uniformity in the great operations of nature, our reasonings would be baseless, our conclusions a dream. We can only read the Past as connected with the Present, and premise of the Future from what is now taking place around us.

Destroy this belief in the continuous operation of natural law, and appeal to "revolutions" and "cataclysms," and you present a world of disorder, a Creator without a plan, and the human reason striving in vain to elaborate a system from phenomena over which no system prevails. Establish this belief, and the geologist feels he is dealing with a prescient plan whose past ever bears certain appreciable relations to its present; and in tracing the development of that plan, he is animated by the high hope of ultimately attaining to some conception, however faint, of the divine idea of its Creator. And it is in this spirit of procedure that he has subdivided the strata of the earth's crust into "systems," and "groups," and "series"—each system being but the sediments of the lakes and seas of a certain period, and characterised, of course, by its own peculiar fossils, as evidence of the life that prevailed during the time of its formation. And the reason is obvious: as land and sea have often changed places—the former at one time more insular, at another more continental; now sitting low and moist in the water, now elevated into lofty and arid regions; subjected at each change to diversity of colder or warmer ocean-currents, to new sets of winds, rains, and other climatal conditions—each period must necessarily have stamped its own impress on vegetable and animal life; and so it happens that the great rock-formations (the only records of the world's history) are each characterised by its own peculiar fossils, or facies of animated existence. Thus, when tabulated, these systems and groups present the following chronological arrangement:—

ROCK-SYSTEMS.	LIFE-PERIODS.	
POST-TERTIARY, ...	} CAINOZOIC (*Recent Life*), ...	NEOZOIC CYCLE.
TERTIARY, ...		
CRETACEOUS, ...	} MESOZOIC (*Middle Life*), ...	
OOLITIC, ...		
TRIASSIC (Upper New Red), ...		
PERMIAN (Lower New Red), ...	} PALÆOZOIC (*Ancient Life*),	PALÆOZOIC CYCLE.
CARBONIFEROUS, ...		
DEVONIAN (Old Red), ...		
SILURIAN, ...		
CAMBRIAN, ...		
METAMORPHIC, ...	HYPOZOIC (*Under Life*), ...	

Such are the main stages into which geologists have arranged the stratified crust of the globe—the great chapters, as it were, of world-history, whose strata, like the leaves of a mighty volume, are indelibly stamped with the forms and characters of extinct vitality. As in human history we speak of the times of Ninevites, Egyptians, Greeks, and Romans, so in geology we refer to Silurian, Devonian, Carboniferous, and other systems; and as Ninevites and Egyptians present a certain similarity or *facies* of civilisation, and Greeks and Romans another, so we unite certain systems, having features in common, into Palæozoic, Mesozoic, and Cainozoic epochs. As to the Time represented by these groups and systems, we have at present no means of determining; but, gauging the past by the present rate of geological change, the amount must be immense, and we could no more form an idea of its aggregate—even could we express it in years and centuries—than we can form a conception of the distances that separate our globe from the remoter stars of the universe. Enough for us, in the mean time, to be convinced of the vastness of its relative portions, and to fix with certainty the order of their occurrence. As in human history it is ever more important to determine the true sequence and connection of events than to be curious about the minutiæ of dates, so in geology it is far more satisfactory to discover the order in time than to indulge in surmises about the expression of its duration in years and centuries. It is surely of higher value to be able to determine the relative ages of two contiguous deposits, the contemporaneity of others widely apart, and the kind and character of life they respectively imbed, than to perplex ourselves with vague hypotheses as to the number of years that have passed since the date of their deposit. And yet even for this, too, the time will undoubtedly arrive! Geological events are the orderly results of natural laws; laws are as fixed in their *times* as in their *modes* of action; and while the Creator has permitted the human intellect to investigate and determine the one, we may rest assured that the same intellect is yet destined to discover the amount and duration of the other. In the mean time, all that geology attempts is to arrange the formation of the earth's crust into so many provisional stages—each stage representing an indefinite amount of time, but embracing such stratified deposits as indicate a contemporaneity of origin, and are characterised by a general similarity of organic remains. In this case, each stage represents the sediments of a certain period, and is necessarily characterised by its own peculiar fossils—every change of sea and land not only giving rise to new sediments, but to altered conditions of vital existence, that are inevitably followed by a modification of the flora and fauna. And summing up the whole, we are presented with the outline, at least, of a grand and continuous evolution of vitality. Here there may be local imperfections in the record—there the characters may be fragmentary and obscure; but in the main the broad features of world-history are sufficiently obvious, and these systems and formations (provisional as they may be) enable the geologist to give intelligible expression to the line and order of occurrence.

Proceeding upon the basis of this arrangement, let us now inquire into the nature of the Plants and Animals preserved in these successive formations. Were they constructed on the same plan, and destined to perform analogous functions in the economy of nature, with those that now live and flourish around us? Or if differing in type, what the amount of that difference, and the presumable function which that difference implies? If race after race has come and departed, what the conditions that accompanied their advent, and what the causes which apparently lead to their extinction? Do the simpler and lowlier forms always precede the higher and more complex; and does the introduction of any family in point of time harmonise with its place in the scale of organisation? Does the extinction of species appear to be, in every case, the result of a change in external conditions; or may not species, like individuals, have a term assigned to their existence from the beginning? If race after race follow each other in order of organisation, what countenance does this give to the theory of self-development? Is there, as far as palæontology can discover, any foundation whatever for the belief in a progressive transmutation of species, by which the lower gives birth to the higher; or does geology not rather establish the conviction of independent creations as time rolled on and new conditions were prepared for their reception? Seeing that *physical* phenomena invariably take place under the orderly operations of natural laws, are we, in the spirit of sound philosophy, entitled to assume for *vital* phenomena any other mode of occurrence? In all other reasonings are we to adopt the inductive method, and in the solitary instance of LIFE—its incomings and outgoings—are we to forsake this course as impotent and unavailing, and appeal to the direct and miraculous interference of Creative Power? These, and numerous analogous questions, present themselves to the palæontologist; and if in human history chronologers are often disagreed as to times and incidents so recent as those that come within the range of a few thousand years, if ethnologists have failed to trace with certainty the relationship of the few varieties of our own race, and antiquarians be only beginning to decipher the phases of certain extinct civilisations, what marvel need it be that geologists are not yet as one as to events for which time has no dates, save "cycles" and "systems," or that they should be occasionally unable to discover the nature and functions of creatures whose remains are so fragmentary, and to whom existing nature offers not a single specific identity? And yet, as we shall afterwards see, geological belief is much more uniform than is generally supposed; and, founding on this belief, palæontology has been enabled, within the brief space of half a century, to establish a history of the world's Past Life, more marvellous by far than the fabled creatures of romance, and yet so true that he who remains in ignorance of its facts can never hope to attain to a satisfactory knowledge of the scheme of life that at present surrounds us.

THE FAR PAST.

PALÆOZOIC SYSTEMS—THE CAMBRIAN, SILURIAN, DEVONIAN, CARBONIFEROUS, AND PERMIAN.

ON glancing over the existing forms of the vegetable and animal kingdoms, struck as we may be at first by their wondrous variety and complexity, we gradually begin to detect innumerable affinities that link one family to another, and at length perceive that one plan and purpose runs throughout the whole. In like manner, when we turn to the still stranger and more complicated forms of the Past, and blend them with those of the Present—varied and endless as the details may appear—they gradually coalesce into one unbroken sequence of design, from the morning that first dawned on infant life, to the sunset that closed around us but a few hours ago. Without this uniformity in purpose and design, the study of nature would be impossible: we can only reason respecting the past from our knowledge of the present, and predict of the future from what is now taking place around us. And here at the outset we must specially guard against the misconception that in the Past Life of the globe we are to meet with anything that is monstrous or abnormal. As in the physical world we have no evidence of the operation of "aberrant" or "cataclysmal" or "revolutionary" forces, so in the vital world philosophy cannot point its finger to a single instance of the abnormal. The "Antediluvian" and "Pre-Adamite monsters," of which we occasionally hear, are the mere creations of the platform orator, who would rather excite the marvellous for the chance of a little applause, than appeal to the reason of his audience by a simple statement of the truth as it occurs in nature. And yet, after all, the works of God are in themselves sufficiently wondrous to arrest the attention, and never more so than when arranged in that simplicity and perfection of design which it is the aim of legitimate science to detect, and the pride of the philosopher to explain.

In treating, then, of the Extinct Life of the globe, it shall be our aim to assimilate its forms, as far as the facts will permit, with those still living around us; to assign to them their places in the scale of being; to note their incomings and outgoings in point of time; and, above all, to discover their functions in the great economy of nature. Important as facts and specific distinctions are to the botanist and zoologist, the discovery of the functions and ultimate design of being is, to our apprehension, a more exalted pursuit;—so true is it (in the impressive words of Coleridge) that "a man may be a chaos of facts, and yet lack the knowledge that God is a God of order." As the establishment of Law appears to be the highest effort of creative energy, so the expression of that law must ever constitute the noblest attainment of created intelligence. And this law is operating everywhere. The force that directs the drifting of a grain of sand is as fixed as that which guides the revolution of a planet; the tiniest blade of grass that turns itself to the sun is but obeying the same law that regulates the growth of the lordliest oak; and the monad, invisible to the naked eye, is the creature of instincts and appetites as imperative as those that impel the actions of man. Nay, not a shower that falls, nor a breeze that blows

—fickle and uncertain as these may seem—but is the result, immediate and remote, of Law, could we only grasp the multifarious conditions that are connected with its production. In tracing, then, the Flora and Fauna of successive epochs, as far as the limits of a popular sketch will permit, we can only indicate a few of their more prominent features and the laws that seem to bear on their development; and yet, restricted as these limits are, enough, we trust, will be indicated to arrest the attention and to arouse the interest in the further prosecution of a subject that stands second to none on the roll of human acquirements. And, after all, it is better to be imbued with the right spirit of research, and to be impressed with the conviction of the universality and uniformity of natural law, than to have the mind bewildered with details which it cannot connect, and for whose occurrence in nature it is altogether unable to render a reason.

And, first, we enter on what has been termed the PALÆOZOIC or "Ancient-Life" period of the world—a period embracing the Silurian, Devonian, Carboniferous, and Permian formations, and characterised, *as far as geological evidence goes*, by the almost total absence of a dicotyledonous flora, by a preponderance of invertebrate life, and by the general absence of the higher vertebrata, as reptiles, birds, and mammals. The lowest in rank seem the earliest in time; and so in this primeval epoch, cryptogams and cold-blooded water-breathers become the leading manifestations of vitality. The strata lying beneath the Palæozoic (as will be seen by a reference to the Geological Record) have been termed the AZOIC or "void of life;" but, more correctly and philosophically, the HYPOZOIC, which merely indicates their position "beneath" the fossiliferous strata, and that without asserting them to be wholly destitute of organic remains. So far as our present purpose is concerned, it matters little which term is adopted, so long as we bear in mind that up to the present day they have yielded no traces of life, and are to all intents and purposes truly Azoic. That the Crystalline or Metamorphic strata, termed *clay-slate*, *mica-schist*, and *gneiss*, were at one time the clayey, sandy, and limy deposits of seas and estuaries, is at once admitted by every competent geologist; and that if these seas contained life, those strata must have imbedded its remains. But then, these deposits have, since their solidification into rock, been subjected to thermal, chemical, electrical, and other agencies, to such a degree that they have been converted, or *metamorphosed*, into crystalline masses, and every trace of life has been obliterated from their structure. No doubt it has been ingeniously suggested that the occurrence in metamorphic rocks of sulphuret of iron, of phosphate of lime, bituminous springs, and other similar products, gives evidence of the presence of organic bodies, through the medium of whose decay such compounds were eliminated. On the other hand, experimentalists equally ingenious have assigned to these products a purely chemical origin; and, even if they could not, the geologist would be little aided by a contrary hypothesis, so long as he had no trace of organic form or texture to guide him in his deductions.

To the palæontologist, therefore, the CAMBRIAN period, with its obscure and scattered zoophytes, trilobites, and shells, becomes the so-called "Dawn of Life." He knows of nothing beyond this primordial zone, and the spirit of true philosophy forbids him to substitute conjecture for fact, or hypothesis for reality. It may gratify the cosmogonist to fashion a glowing globe by the condensation of nebular masses, to cool by radiation a solid crust on the glowing orb, and, after ages of chaotic confusion, to plant the germ of life on some sunny and serene spot;—it may charm the materialist to claim for Life the eternity he does for Matter, by referring to a metamorphism which is continuously obliterating the fossils in the deepest seated rocks; but the palæontologist is debarred from such reveries, and is bound down by a rigid chain of facts as they occur in nature. He has traced life so early as the Cambrian slates; should it be detected still lower, he is ready to accept it. To him, in the mean time, the Metamorphic schists are a *tabula rasa;* the Cambrian slates form his furthest verge and boundary; and the spirit of induction restrains him within its limits. And, after all, fossil evidence itself is greatly in favour of the view, that we have here attained, or all but attained, the furthest limit of life. We see it increasing and spreading into higher and higher forms as we ascend in the geological scale, and decreasing and narrowing into lowlier forms as we descend: numerically the forms are fewer, physiologically they become less important; and it is but fair induction to believe that in the few scattered forms of Cambria we have all but reached the zero of organic existence.*

From the Cambrian the palæontologist passes into the Silurian age—a period characterised by its lowly sea-weeds and doubtful traces of land plants—by genera and species of protozoan, radiate, molluscoid, molluscan, and articulate types, but by few, if any, even of the lowest vertebrate order. Its strata consist of shales, sandstones, conglomerates, and limestones—the solidified muds, sands,

* It is right to mention, however, that the tendency of recent discovery is to carry the traces of life further and further back among these slaty and semi-crystalline strata. The detection of new graptolites and trilobites in the schists of Bray Head, Skiddaw, Bohemia, and North America, is a fact too significant to be overlooked in geological speculation.

pebbles, and coral-growths of seas and estuaries. It is customary for a certain class of geologists to talk of "the deep, turbid, and shoreless seas" of the Silurian epoch, as if the globe was then enveloped by one dreary monotony of ocean. Do such generalisers ever for a moment think that such a vast thickness of sediments could never have been produced without the existence of broad lands from which they were transported by rivers, or of sea-shores from which they were abraded by waves and tidal currents? Could conglomerates be formed without wave-exposed beaches, sands without open sea-shores, or could shells that are truly littoral, and corals that flourish only from twenty to sixty fathoms, have existed without water of limited depth for their development? The eye of the trilobite would have been useless in a turbid ocean; a turbid ocean would have been death to the growth of corals; worm-burrowed, ripple-marked, and rain-pitted sandstones could have been formed only on shores exposed to the alternate ebb and flow of the tide; and conglomerates are merely the broken-down and water-worn fragments of an older rocky shore. In fine, there is not a single feature in the rocks of the Silurian period which might not take place in the ocean of our own day. The existence of deeper and shallower seas—of waves, currents, tides—of lands, shores, and rivers—of sunlight, and rains, and winds—are as clearly impressed on its strata as they are upon those of every other geological epoch. It differs alone in the geographical distribution of its sea and land—the greater insularity, perhaps, of the land-masses—their consequent climatology—and the specific characters of its plants and animals; though, knowing the wide extent of its deposits (and they occur alike in the continents of the Old and New World, in the northern and in the southern hemisphere), geology is not yet in a position to map with accuracy the geography of the

SILURIAN ERA.

period, nor to define with certainty the external conditions to which its flora and fauna would be necessarily subjected.

When we turn to its biological aspects, the outline, though far from complete, is at least, as far as it goes, homogeneous and intelligible. *Fucoids* or fucus-like sea-weeds, some carbonaceous fragments of unknown stems, *spore-like organisms*, apparently from land plants, and a few *lepidodendroid* twigs that may have belonged to some ancient form of club-moss, are nearly all we know of the silurian Flora; though, judging from the extent of anthracite deposits in various regions, vegetation (aquatic and terrestrial) must in certain centres have existed in some

FRAGMENTS OF SILURIAN FLORA.

1, 2, Fucoids—Cruziana and Chondrites (?); 3, 4, Lycopodites—Lepidodendroid twigs from the Upper Silurians of Lanarkshire.

exuberance. On the whole, the silurian Flora is of a very lowly character, and its scanty fragments find their nearest affinities in the sea-weeds, liver-worts, and club-mosses of existing nature. Of course, the imperfection of the geological record is fully and frankly admitted, for it cannot be supposed that in strata so eminently marine, we are likely to discover more than the merest indication of a terrestrial vegetation. Still we can only reason from what we know, and shape our inferences by the results of our observation.

When we turn to the Fauna of the system, we find the record much more complete and legible. We are presented with *infusorial* organisms from its shales; *graptolites* or sertularian-like zoophytes in inconceivable numbers; *corals* of

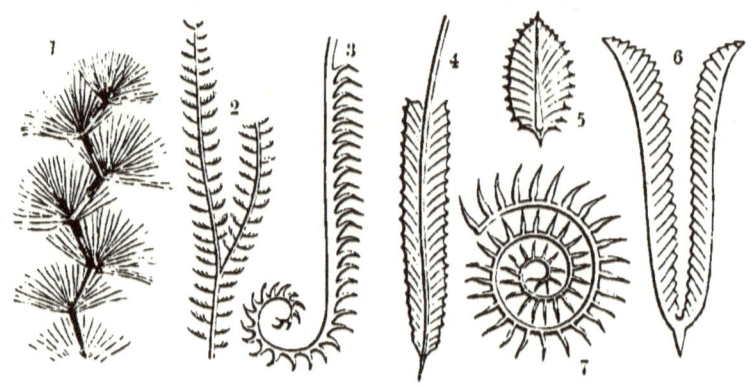

SILURIAN HYDROZOA AND BRYOZOA.

1. Oldhamia; 2. Protovirgularia; 3. Graptolites; 4, 5. Diplograpsus 6. Didymograpsus; 7. Rastrites.

many genera and species; *encrinites* of various forms; *star-fishes*, independent and free-floating; and sea-urchin-like *cystideæ*, attached to the sea-bottom by their jointed foot-stalks. In *molluscan* life we have representatives of every order—brachiopods, acephalans, gasteropods, pteropods, and cephalopods — vegetable-feeders thronging the shores, carnivorous orders in the open sea, and infusorial-feeders in the deeper waters. The great preponderance of brachiopods over acephalans and gasteropods is one of the most noticeable features in the molluscan life of the period —a feature now reversed, seeing that acephalans and gasteropods are the predominating forms in existing waters.* In the articulate division we have numerous *annelid* markings—the trails and burrows of sea-worms; the calcareous crusts and shell-like cases of *serpulæ* and *spirorbes;* and a vast and characteristic display of *trilobites* (three-lobed), a form of crustacean almost restricted to the period; together with the larger and higher forms of *eurypterites* (broadfins—in allusion to their paddle-like swimming limbs). These

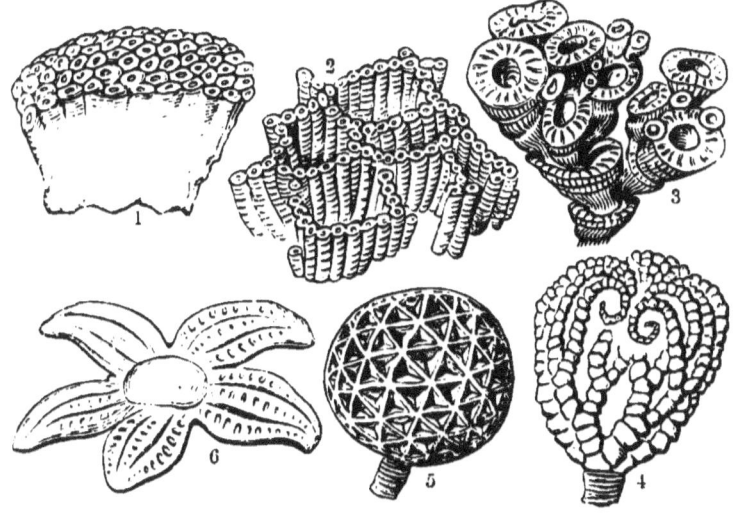

SILURIAN CORALS AND ECHINODERMS.
Halcolites; 2, Catenipora; 3, Cyathophyllum; 4, Taxocrinus; 5, Cystidea; 6, Palæaster

trilobites, along with some smaller bivalved forms of crustacea, have been long and familiarly known; but the euryp-

* We abstain, in this as in other instances of comparison, from numerical tabulations, as every year of further discovery and nicer discrimination of species disturbs, if not destroys, the value of such statistics. Not many years ago the Brachiopoda were supposed to be on the very verge of extinction, and yet the application of the dredge to deeper waters has revealed the existence of nearly a dozen genera in modern seas. Every year, too, discovery adds some new form to our lists of fossils, while former lists of so-called *species*—Continental, British, and American—are being examined with more rigorous care, and reduced to their proper value.

terites are a comparatively recent discovery in the higher

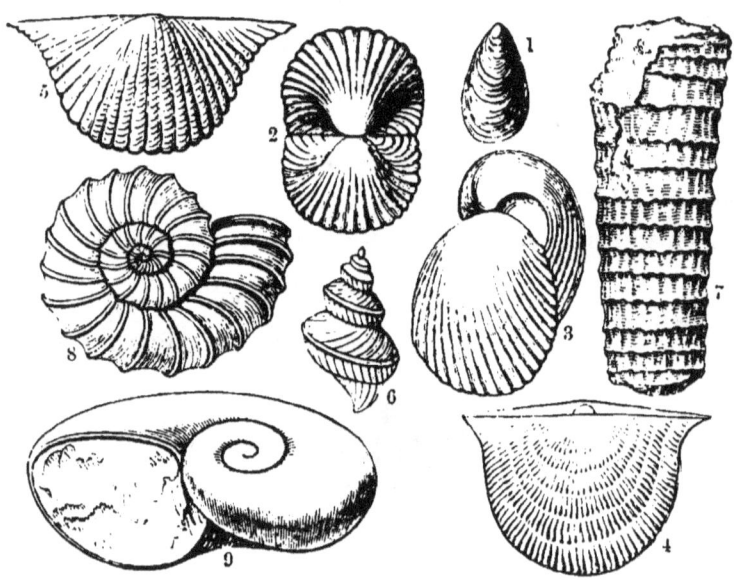

SILURIAN MOLLUSCA

1. Lingula; 2, Rhynconella; 3, Pentamerus; 4, Strephomena; 5, Spirifer; 6, Murchisonia 7, Orthoceras; 8, Lituites; 9, Maclurea.

beds of the system, and two of the most abundant genera

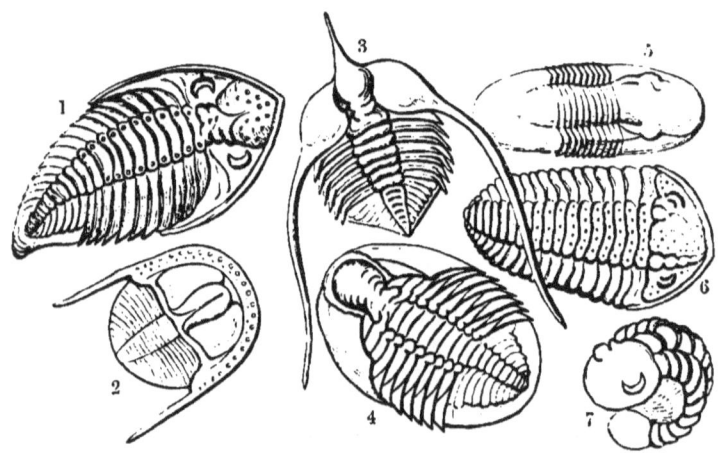

SILURIAN CRUSTACEA—TRILOBITES.

1, Phacops; 2, Trinucleus; 3, Ampyx; 4, Ogygia; 5, Ilænus; 6, Calymene; 7. Calymene coiled up.

are here for the first time restored with something like accuracy and life-like proportions. As already hinted, the remains of *fishes* are but sparingly found in the uppermost beds of the system; and it is still an open question with

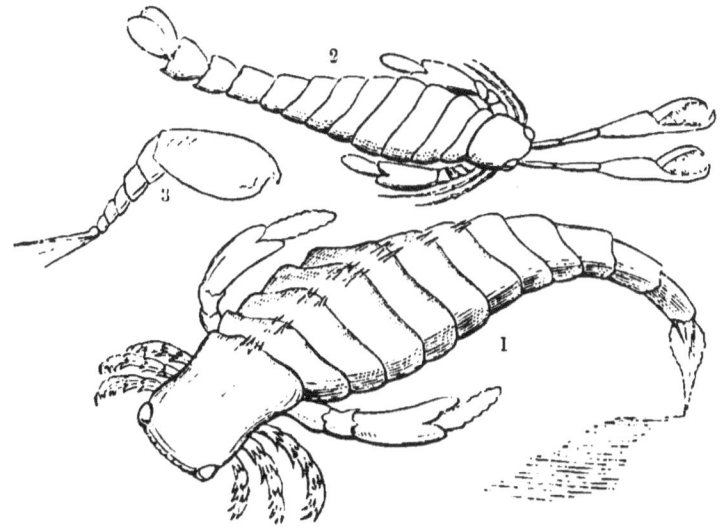

SILURIAN CRUSTACEA—EURYPTERIDES.

1, Pterygotus Acuminatus; 2, P. Bilobus; 3, Ceratiocaris (bivalved Crustacean). From the Upper Silurian or Passage Beds of Lanarkshire.

geologists, whether these are to be viewed as marking the close of the silurian or the dawn of the devonian epoch. For our own part, we accept them as part and parcel of the silurian fauna; and though negative evidence forbids us in the meanwhile to enter on our lists the remains of insects, reptiles, birds, and mammals, there is nothing that militates against the likelihood of their occurrence. On the contrary, all analogy favours the supposition that the great types of life—radiate, molluscan, articulate, and vertebrate—were from the beginning contemporaneous on our globe, and that it is to the minor modifications of the type, and not to the type itself, we are to look for that gradation and progress which marks the geological periods. In this opinion we are further fortified by the decidedly expressed conviction of one of the ablest investigators of the present age. "However much naturalists may still differ in their views regarding the origin, the gradation, and the affinities of animals," says Professor Agassiz in his Essay on Classification, "they now all know that neither radiata, nor molluscs, nor articulata have any priority one over the other as to the time of their first appearance upon earth; and that, though some still maintain that vertebrata originated somewhat later, it is universally conceded that they were already in existence towards the end of the first great epoch in the history of our globe. I think it would not be difficult to show, upon physiological grounds, that their presence upon earth dates from as early a period as any of the three other great types of the animal kingdom, since fishes exist wherever radiata, molluscs, and articulata are found together, and the plan of structure of these four great types constitutes a system intimately connected in its very essence. Moreover, for the last twenty years every extensive investigation among the oldest fossiliferous rocks has carried the origin of vertebrata step by step farther back; so that, whatever may be the final solution of this vexed question, so much is already established by innumerable facts, that the idea of a gradual succession of radiata, molluscs, articulata, and vertebrata, is for ever out of the question."

Here, then, in the Silurian system we find nothing abnormal or marvellous! Its sediments tell of seas whose shores, in favourable localities, were clad with weeds, and whose waters were thronged with zoophytes, star-fishes, sea-urchins, shell-fish, and crustacea. Plant-feeder and animal-feeder start simultaneously in the race of life; and it requires no great stretch of fancy to repeople Silurian waters, busy and joyous on a summer's eve as the tribes that throng the existing ocean. The life-forms of the period are, in their kind, neither larger nor smaller, neither less perfect nor less complex, than those of the current era. From the beginning, and simultaneously, species and genera and orders assume their distinctive characters; there are no transitional forms (in the ordinary sense of the term) through which we can trace the development of the higher from the lower; each species takes its place from the beginning, and varies only within a certain defined limit; while the whole, obeying the impulses and instincts of life, subserve with unerring certainty the creational functions they were destined to perform.

We now pass from the Silurian to the Old Red sandstone, or, as it is now more frequently termed, the *Devonian* epoch. And here, in its sandy and pebbly deposits, we find more decided evidence of frequent alternations of sea and land, of broad shallow bays, and long reaches of shingle-covered shores. Much of the silurian deep sea had been upheaved into dry land; the former islands and continents had received new configurations and altitudes; and the seas so changed must have been subject to the influences of other tides and currents. We have also clearer evidence of estuarine and lake areas; and were this the place to enter on questions of physical geology, testimony is not wanting to prove the existence in certain regions of a cold or glacial climate.* All this implies numerous modifications

* Whoever has examined the bouldery conglomerates of the Scottish Old Red, with their large irregular blocks, their peculiar unassorted aggregation, the nature of the cementing matrix, and the frequent "nestings" or interlaminated patches of fine argillaceous sandstone, must have had suggested to his mind the idea of ice-action. And this notion must have been strengthened when he turned to the sandstones, and found them imbedding angular fragments of rock, shale, and even clay, which could scarcely have suffered transport unless enclosed in drifting ice-floes. The paucity of life in certain areas seems also a further

of external conditions, under the influences of which many of the silurian genera and species became extinct, and other races were introduced specially adapted to the physical circumstances by which they were surrounded.

In the Vegetable World we have now a greater exuberance and variety of *fucoids* or sea-weeds; marsh plants, apparently related to the equisetums, reeds, and rushes;

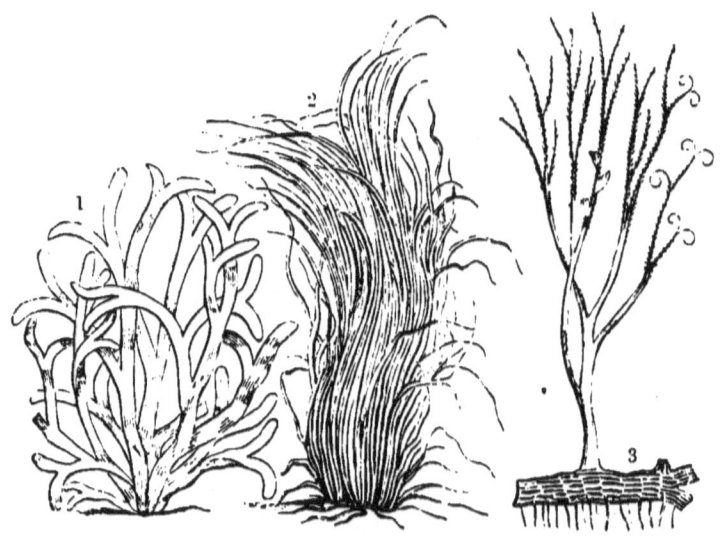

FRAGMENTS OF DEVONIAN FLORA.

1, Fucoid (Roxburghshire); 2, Zosterites (Forfarshire); 3, Psilophyton (Canada), Dawson

and unmistakable evidence of a terrestrial flora of no feeble growth. Ferns of rare beauty (*adiantites*), club-moss-like stems of gigantic growth (*lepidodendra*), and fruit cones (*lepidostrobus*), are by no means uncommon, and every year is adding some new feature to a flora which a dozen years ago was set down as having no existence.

corroboration of the idea of glacial influences—an hypothesis which seems at first sight extremely probable, though requiring for its final demonstration a much more protracted and careful examination than the several phenomena have yet received from geologists.

Even in some more favoured spots, like Point Gaspé, in Canada, thin seams of bituminous coal are interlaminated with the plant-yielding shales and sandstones, thus giving further proof that the old red period had its areas of fertility and areas of dwarfish sterility—regions where climatic influences were mild and genial, and others where they were rigorous and destructive of vegetation.

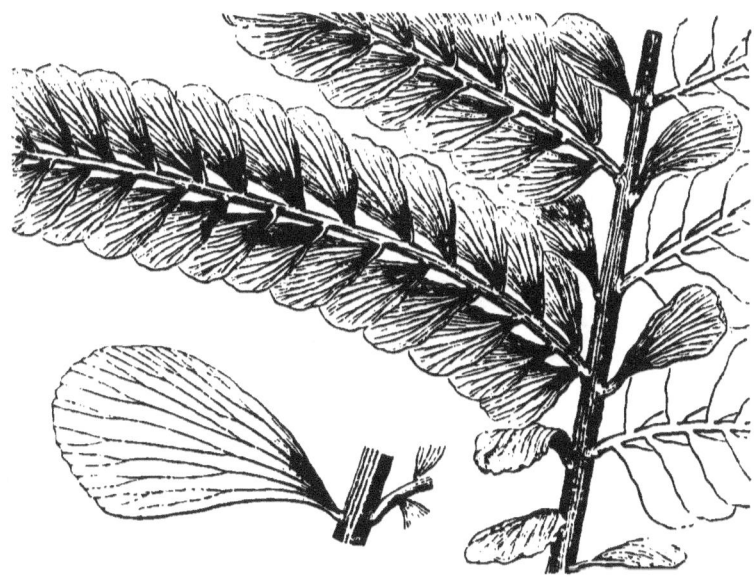

Adiantites Hibernicus—Yellow Sandstone Series of Ireland.

In the Animal World we have still the same numerical abundance of *zoophytes*, of brachiopod, gasteropod, and cephalopod *molluscs;* but the *graptolites*, which flourished in such profusion in the muddy bottoms of the silurian seas, have become extinct; the *trilobites*, whose species were then numbered by hundreds, are reduced to a dozen or two; while the larger crustacean forms of *eurypterus*, *pterygotus*, and *stylonurus*, then merely appearing, now flourish in great force; and fishes of various families swarm in vast profusion. Gigantic *annelids*,* large as a man's arm, throng the sandy shore, leaving, like the lobworm, their frequent casts

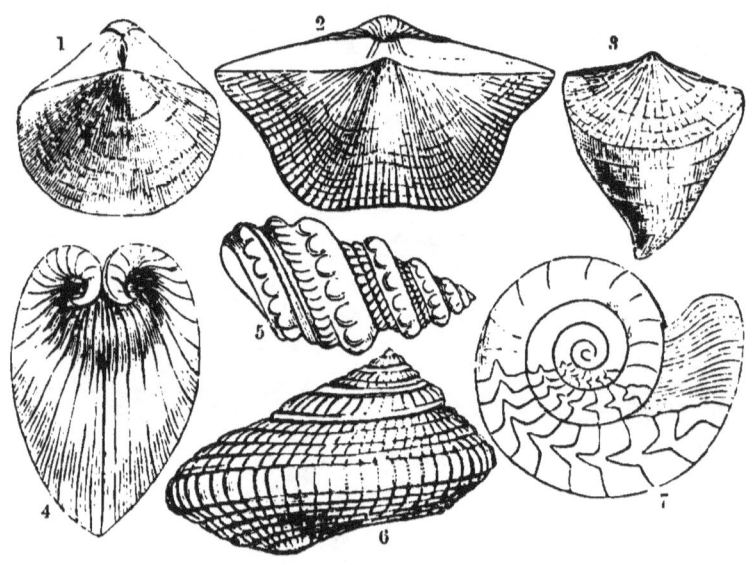

DEVONIAN MOLLUSCA.

1, Stringocephalus; 2, Spirifera; 3, Calceola; 4, Megalodon; 5, Murchisonia; 6, Pleurotomaria; 7, Clymenia.

and burrows, while smaller species and wandering crustacea thickly track the rippled, and rain-pitted, and sun-cracked surface with their devious courses. Reptiles also, for the first time, come into notice—there being no great order in existing nature unrepresented, save insects, birds, and mammals.

The most noticeable feature in the fauna is, perhaps, the large crustacea, the curiously encased fishes, and the

* These so-called *annelid burrows*, which occur alike in the lowest Old Red of Forfar, and in the upper beds of Roxburgh, are deserving of a closer examination than they have yet received from the palæontologist. Some of them are so large and of such curious internal configuration, that one is tempted to inquire whether pterygotus and his allies did not occasionally burrow their abdominal segments in the sandy mud, and there, pincers at rest, watch for their passing prey.

occurrence of reptiles—the earliest of their class positively known to geology, if British observers be not mistaken as to the relations of the strata in which their remains have been detected. Of these crustaceans, found chiefly as yet in Forfarshire and Hereford, we know too little to assign to them their exact place in existing classification; but if concentration and specialisation of organs are to be tests of higher and lower, then we are compelled to place them rather low in the class to which they belong. Like many extinct forms, the *eurypterites* partake of the characters of several adjacent orders, and thus, like the dis-

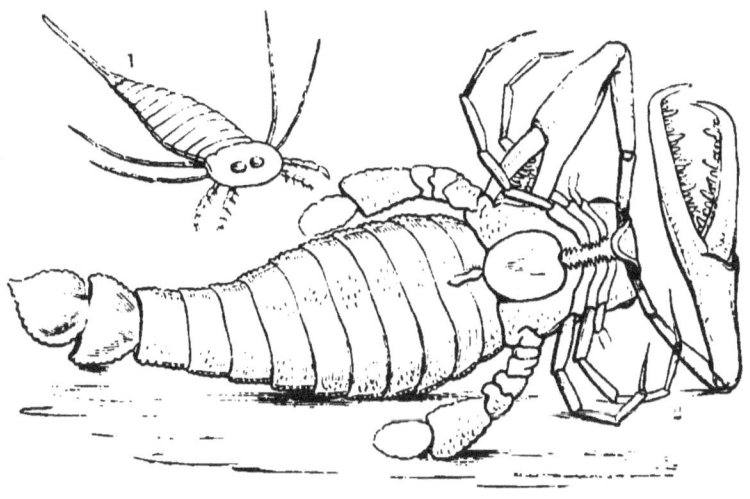

DEVONIAN CRUSTACEA
1, Stylonurus Powriei; 2, Pterygotus Anglicus (ventral aspect). From the Lower Old Red of Forfarshire.

covered portions of some ancient mosaic work, they fill up the gaps, and bring out more clearly the continuity of the design that runs throughout the whole. King-crab-like in their carapace and organs of mastication, lobster-like in their prolonged and segmented bodies, furnished with broad paddle-like swimming limbs, and frequently with huge prehensile claws, they present the zoologist with an entirely distinct family (Eurypteridæ), if not with the elements of a new and separate legion. Some of the species are of great size—three, four, and six feet in length—and seem to have been the scavengers of their period, living on the lower forms and garbage of the sea-shore; and so it happens that wherever they are found entire fishes are rare, though their heads, fin-spines, and other unmanageable portions occur in abundance. Another curious feature in connection with these crustacea, and occurring in the same beds, is an immense number of dark-coloured patches of spawn-like organisms, which are now pretty generally regarded as the egg-packets of eurypterus and pterygotus. Compressed and flattened, the ova appear less

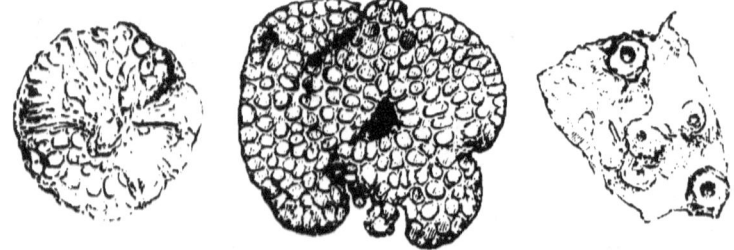

Parka Decipiens—Supposed Spawn or Egg-packets of Crustacea.

or more in concentric arrangement, and every appearance favours the idea of their crustacean origin, unless, perhaps, their great abundance, which has suggested to some the possibility of their being the berries or carpels of some unknown plant. The egg-packet theory is now the most prevalent, and, admitting its truth, the widespread abundance of these remains increases beyond expression our notions of the exuberance of crustacean life within certain areas of the old red sandstone.

The Fishes of the period are also peculiar, inasmuch as many of them are encased in bony plates, or covered with hard enamelled scales; are frequently furnished with fin-spines or external defences; and are, many of them, of forms so widely differing from those of existing seas, that they have not unfrequently been mistaken for reptiles, for crustaceans, or even for huge water-beetles! And yet, when closely examined, and their affinities made out, there is nothing about them either abnormal or nondescript. The more familiar forms are the *cephalaspis*, or "buckler-head," so called from the shield-like shape of the bony head-plate, which consists of a single piece; the *pterichthys*, or "wing-fish," having the body encased in a box-like covering of bony plates, and furnished with two wing-like appendages for swimming; the *coccosteus*, or "berry-

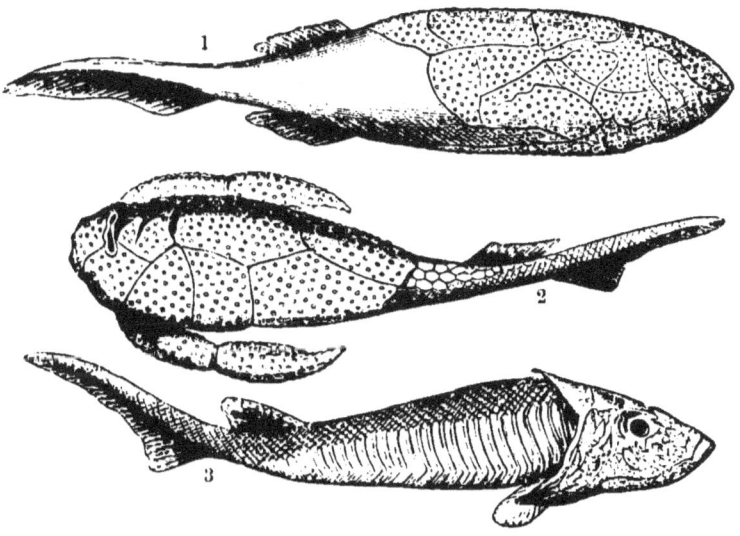

1, Coccosteus; 2, Pterichthys; 3, Cephalaspis.

bone," similarly encased, and having the surface of the plates covered with minute berry-like tubercles; the *acanthodians* (diplacanth, cheiracanth, &c.), having for the most part their fins armed and supported by bony spines; the *dipterus*, or "double-fin;" the *osteolepis*, or "bony-scale;" the *asterolepis*, or "star-scale;" and the *holoptychius*, or "all-wrinkle," so

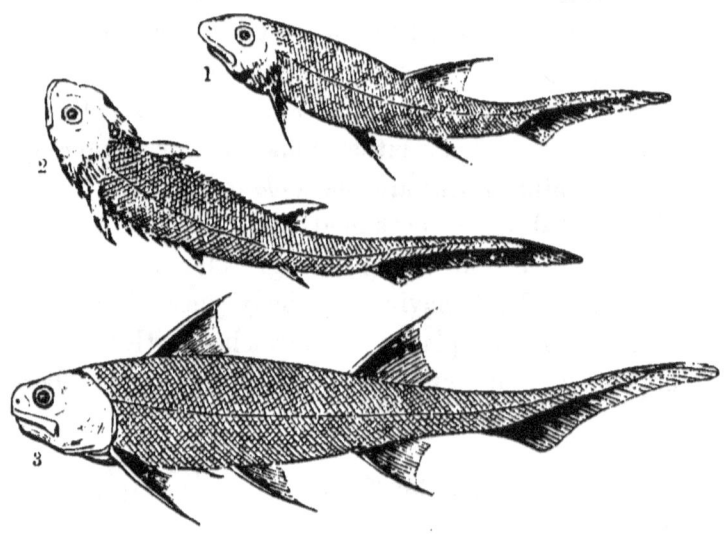

1, Acanthodes; 2, Climatius; 3, Diplacanthus.—Forfarshire.

called from the wrinkle-like sculpturing that adorns its large enamelled scales. The majority of these fishes are small, or of moderate size; and even the largest of them, the holoptychius and asterolepis, do not greatly, if at all, exceed the dimensions of a full-grown cod-fish. Nor would they startle by their forms, were they recalled to take their place among existing fishes. The little armed bull-head of our own shores is encased in as marvellous, and even more highly ornamented armour than the cephalaspis; the ostracion, or trunk-fish of the Indian ocean, is encased in a bony box, as curiously fabricated as that of the pterichthys or coccosteus; the spines of the balistes and sea-snipe are as formidable weapons as the ichthyodorulites of the diplacanth; and the scales of the bony-pike of South America, or the polypterus of the Nile, glitter with enamel, and are as quaintly sculptured as those of the osteolepis or holoptychius of the old red sandstone. Wonderful they are, as all God's works are wonderful! but to dwell, as is too often the case, on these ancient denizens of the deep as something unusually strange and marvellous, is neither the way to forward the interests of science, nor to teach the popular mind a just appreciation of the world that surrounds it.

Turning next to the higher order of reptiles, we have as yet no undoubted instance of their existence, though footprints, bones, teeth, and scutes, unquestionably reptilian, have been detected in the sandstones of Elgin— sandstones till recently regarded as Devonian, but whose relations have lately been questioned; *first*, on account of the obscurity of their stratigraphical relations to the surrounding old red sandstone of the district; and, *second*, on account of the affinity of their reptilian remains to those occurring in true triassic strata. Partaking in the doubt, both on lithological and palæontological grounds, that the place of these Elgin sandstones may yet be found to belong to the dawn of the triassic, and not to the close of the Devonian, epoch, we have transferred the small lizard-like *telerpeton*, and the large " drop-scaled" crocodilian *stagonolepis*, to the newer era—a transposition already approved by some of our leading palæontologists. It is thus that palæontology often corrects the first impressions of physical geology, and in this instance conformably so with all that we know of the reptilian life of the carboniferous era, where forms more lowly and fish-like in their character alone make their appearance. In the mean time, therefore, the existence of reptiles during the old red sandstone epoch must be held as problematical, and palæontology constrained to date their advent with the commencement of the carboniferous era. If it shall be ultimately found that these Elgin sandstones are of true Devonian age, the occurrence of reptiles having such high affinities as lizards and crocodiles, will once more correct the hasty generalisations of limited observation, and teach us how vain it is to dogmatise on the rise and order of life from the imperfect data which geology has yet at her command.

Such is the hurried glance at the life of the Devonian epoch. As yet we are almost in total ignorance of its terrestrial flora and fauna. We are like voyagers to whom some unknown land looms in the distance through the sea-fogs and grey of the morning. Here and there a few gleams of light fall on hill-sides green with ferns and club-mosses; and as the mists roll away we catch a passing glimpse of some river-mouth fringed with reeds and rushes. This, however, is all—the interior is obscured from our vision, and no drift of fruit or forest-growth tells of a higher flora. As we coast along, we almost think we catch the reflection of glacier and icebergs, which would indicate in some regions a sterility and dearth of vegetation; but this may be a delusion, and only the sparkle of the quartzy cliffs that are broken into fragments by the surf that dashes against them. When we turn to the ocean, the view is somewhat nearer and clearer. In the warmer seas, corals of various form and beauty are rearing their reefs; shell-fish of every grade, though not of great numerical abundance, are busy along shore and in mid-water; fishes of widely different forms swarm in shoals —generically few, but individually most numerous; whilst crustaceans of uncouth shape and gigantic growth feed on the tide-borne garbage of the muddy creeks and shallow lagoons. This is all: and much as has been made of it, all reason forbids us to accept it as more than the merest contribution to the biology of the period.

Succeeding the old red sandstone, and much more sharply defined—physically and vitally—comes the great CARBONIFEROUS FORMATION. We have now extensive alterations in the distribution of land and water—shallower seas—larger rivers and estuaries—wide, far-stretching swampy lands; and with these, new ocean-currents, a more genial and equable climate, and, as a concomitant, a more exuberant exhibition, and over wider areas, of vegetable and animal life. In some regions, but by no means over the whole world, the transition from the one period to the other seems to have taken place through convulsive energy; and hence in these regions comparatively few of the forms of the old red sandstone survive, or pass into the carboniferous era. As in every other period, the new forms come slowly and gradually on the stage, attain their "culminating point," or period of greatest variety, size, and numbers, and then gradually or quickly decline, according to the continuity of the conditions by which they are surrounded. In the vegetable world we have now a most exuberant Flora—so exuberant that it is but faintly paralleled by the rankest growth of the tropical jungle. To account for this extraordinary development of plant-life, over such wide and diversely situated regions of the globe, various hypotheses have been offered, such as a larger percentage of carbonic acid gas in the atmosphere—the greater effect of the earth's central heat—change in the earth's axis of rotation, so as to bring the coal-bearing areas within the tropics—and greater eccentricity of the earth's orbit, so as to have brought the globe periodically nearer to the sun's influence; but as we have not in the mean time* a shadow of proof for such abnormal causes, and much evidence to the contrary, we are bound by sound induction to seek for the explanation in

* We say *in the mean time;* for the recurrence of colder and warmer cycles over the northern hemisphere, as evinced by the geological record, is clearly the result of some great cosmical law, depending either on telluric or on solar influences, and, as such, must sooner or later be satisfactorily determined. See Concluding Chapter—"The Law," Section 6.

the then peculiar distribution of sea and land, in the altitude of its shores, in the arrangement of warmer aërial and oceanic currents, and generally in a concentration of these conditions, such as would produce the necessary climate. And, after all—as in the case of the great tertiary elephants and rhinoceroses of Northern Europe, whose representatives are now found only in the tropics—we know too little of the nature of the plants to say under what conditions of climate they would attain their greatest exuberance, though we clearly perceive from their foliage and mode of growth that it was at once equable and continuous.* Generally speaking, we find them resembling equisetums, marsh-grasses, reeds, club-mosses, tree-ferns, and coniferous trees; and these in existing nature attain their maximum development in warm-temperate and subtropical, rather than in equatorial regions. The Wellingtonias of California, and the pines of Norfolk Island, are more gigantic than the largest coniferous tree yet discovered in the coal-measures; the tree-ferns of New Zealand luxuriate in humid and shady spots; the tussack of Falkland Island, and the phormium of New Zealand, show leaves as broad and long as the *poacites* of the carboniferous period; while accumulations of peat-growth are the products of coldly-temperate, rather than of equatorial latitudes. Besides all this, we have coal-beds in other formations—the oolite, the Wealden, and tertiary; and if we are to go in search of abnormal conditions for the production of the one, we must admit the existence of similar causes for the production of the other— an admission, as we shall afterwards see, that would lead to

* It is more than likely, as suggested by the late Robert Brown, that many of the Coal-plants were inhabitants of the swamp and shallow waters—estuarine and marine; and that, rooted in mud, rich in organic matters, and surrounded by water of an equable and genial temperature, they enjoyed the conditions at once of a rapid and of a gigantic growth.

irreconcilable absurdities. The fact is, coal is a necessary product of every period, and is merely the mineralised result of vegetable accumulation — pointing rather to immensity of time than to rapidity of growth as the cause of that accumulation. It is to time, therefore, and to genial equability of climate, rather than to excessive temperature, that we are to look for an explanation of the vegetable masses of the coal period ; and he who would cut short the difficulty by appeals to abnormal conditions, instead of exhausting the possibilities within the scope of natural law, at once does violence to Nature, and retards the progress of legitimate induction.

The vegetation to which we allude consists of *fucoids* and *confervites*, or sea-weeds and confervæ ; of *equisetites*, *hippurites*, and *asterophyllites*, gigantic plants resembling the horse-tails of our swamps and ditches; of innumerable tree-ferns distinguished by the forms and venation of their leaves, as *neuropteris* (nerve-fern), *cyclopteris* (circle-fern), *glossopteris* (tongue-fern), *pecopteris* (comb-fern), *sphenopteris* (wedge-fern), and the like; of fern stems, *caulopteris;* of reed-like plants, *calamites;* of palms,* *palmacites* and *Noeggerathia;* of a vast variety of trees of unknown relationship, as *sigillaria* (fluted bark), *stigmaria* (dotted bark), now known to be the roots of sigillaria, &c., *lepidodendron* (scaly stem), *bothrodendron* (pitted stem), *favularia* (honey-combed bark), and the like ; and of true coniferous trunks

---

* It has been recently questioned, and apparently on good grounds, whether we have certain evidence of the existence of palms during the Carboniferous epoch? The three-cornered fruits (*trigonocarpum*), formerly supposed to be those of palms, are now regarded as those of coniferous plants, which, like the berry of the juniper, was enclosed in a fleshy envelope ; while the broad flabelliform or fan-shaped leaves (*Noeggerathia*) are also considered coniferous and akin to the existing subtropical Salisburia. The so-called palm-stems have always been held as doubtful.

resembling the pine, araucaria, peuce, yew, &c., and hence known as *pinites, araucarites, peucites,* and *taxites*. As

RESTORED ASPECT OF CARBONIFEROUS FLORA
Calamite; Bothrodendron; Equisetes; Asterophyllites; Lepidodendron Cu...
...fern; Sigillaria, with Stigmaria roots; Tyc...

yet, we have only two or three doubtful instances of a dicotyledonous flora—the majority of the preceding forms being monocotyledons and conifers. Occurring as they do in stony and carbonised fragments, their relations are but ill understood; and botanists have as yet contented themselves by pointing out resemblances rather than in establishing true affinities. Whatever their nature, they must have grown in vast luxuriance and variety, clothing every riverside and plain, and spreading over every swamp in one impenetrable jungle,—and this, season after season, and age after age, till their accumulated growth completed the coalbeds now so indispensable to the progress of civilisation. It is customary for a certain class of writers to descant on the "dreary and flowerless monotony" of the vegetation of the Coal period. This, however, is an error. Though all the equisetums and club-mosses and ferns were undoubtedly flowerless, the higher gymnogens and endogens were not so, as we have evidence in the fossil flowers and fruits (*antholites* and *carpolites*), which thickly stud many of the shales; and we have often thought that what was wanting in blossom was more than compensated for by the profusion of light symmetrical, feathery fronds, and by the tall pillar-like stems which rose, each one boldly carved with its own peculiar pattern. The trunks of a modern forest are rough and gnarled; those of the period now under review sprang up like the sculptured shafts of a medieval temple, graceful in proportion, and rich in ornament through the endless repetition of flutings, spirals, zigzags, lozenges, ovals, and other geometrical designs—these designs being the persistent leaf-scars of a vegetation simpler in structure and more primitive in plan.

When we turn to the Animal Life of the period, which belongs almost exclusively to the waters, we find it equally exuberant in numbers and in variety. Life abounds near shore and in the shallow waters—life is rife in the deeper ocean—and over all there is a *facies* or general resemblance that stamps the period as distinct from the old red sandstone that precedes, as from the new red that follows. The seas swarm with *zoophytes* of various families, and reef-building corals (*astræopora, cyathophyllum, clisiophyllum*, and *lithostrotion*) pile up the masses of the mountain

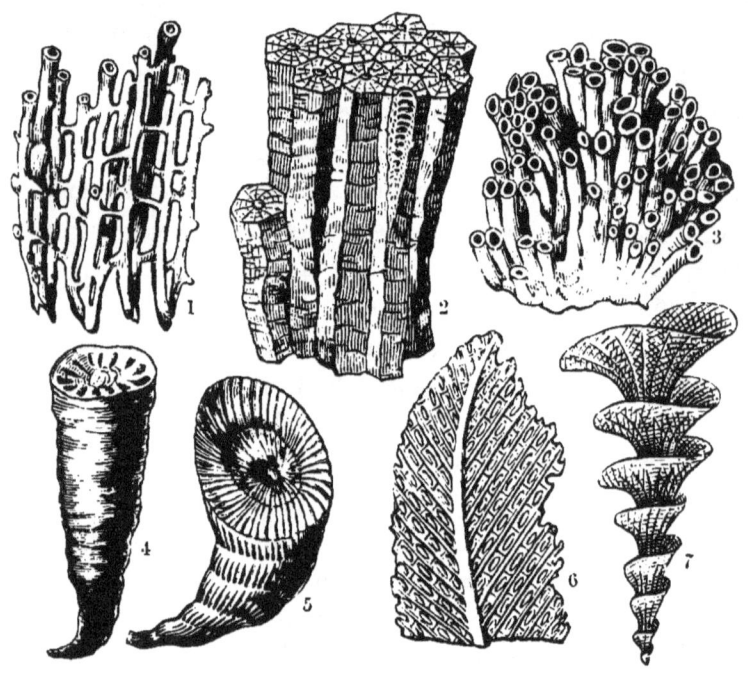

1, Syringopora; 2, Lithostrotion; 3, Aulopora; 4, Amplexus; 5, Clisiophyllum; 6, Ptilopora; 7, Archimedopora.

limestone. Star-fishes (*pentremites*), sea-urchins (*palæchinus* and *archæocidaris*), and *encrinites* of numerous genera and species abound—the latter in such profusion that they now outweigh the zoophytes, and whole strata are composed of their calcareous remains. *Serpulæ* and *spirorbes* attach their sheaths to every available object; sea-worms, like the *arenicola*, leave their tracks and burrows in the sands; and these are also patterned with footprints, pitted with rain-drops, and crested with ripple-marks. In the stagnant lagoons, minute crustaceans, like *cypris* and *cythere*, swarm in myriads; a few species of *trilobite* still

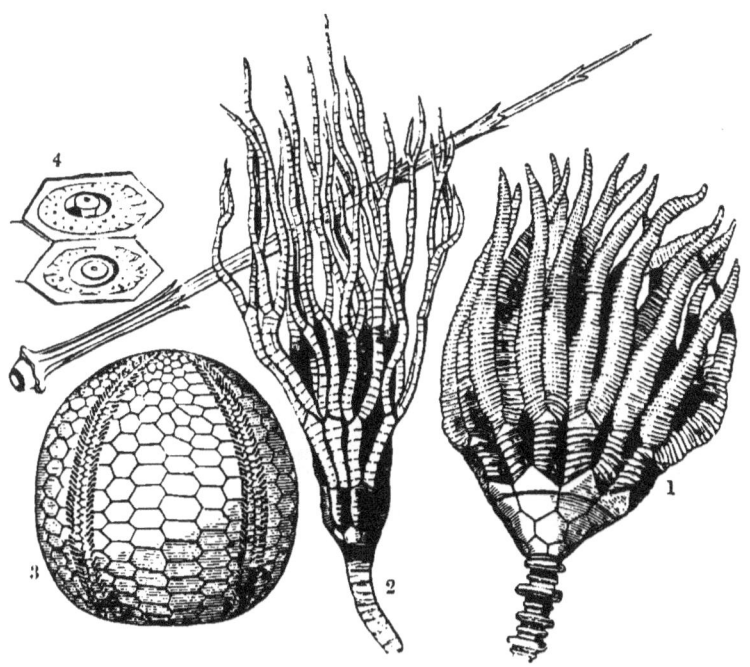

1, Woodocrinus; 2, Cyathocrinus; 3, Palæchinus; 4, Plates and Spine of Archæocidaris.

linger in the muddy creeks; *eurypterites* are on the wane, and forms like the *limulus* or king-crab of the Indian Ocean now make their appearance. For the first time, too, we discover the crusts and wing-cases of beetle-like insects, showing that the vegetation of the period afforded them abundance of food; and that garbage, perhaps of an animal nature, was there also, though we have not been enabled to trace the connection. Every order of molluscan life is busy in the waters—bryozoa, like the *flustra* and *retepora* of our own seas, spread their cells in symmetrical network on dead shells and broken encrinites; brachiopods, like *spirifer* and *productus*, are in deep water; huge nautilus-like cephalopods, *nautilus*, *goniatite*, and *orthoceratite*, in

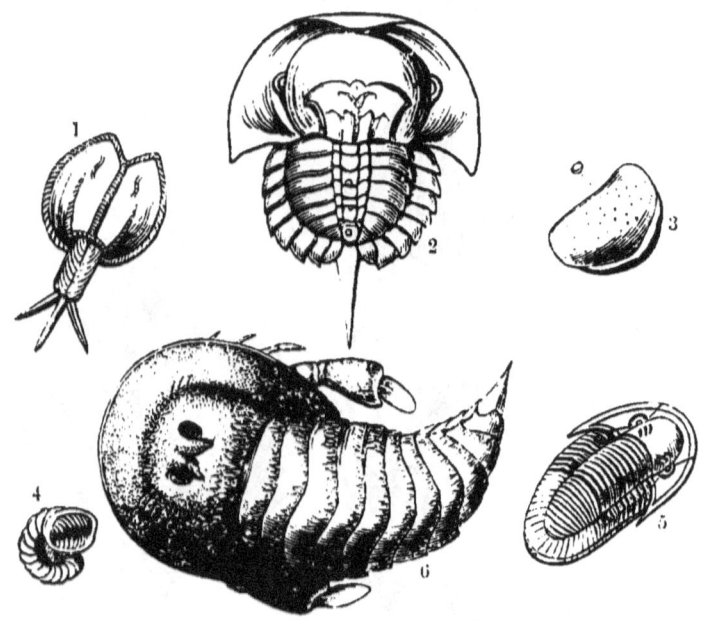

CARBONIFEROUS CRUSTACEA, &c.

1, Dithyrocaris; 2, Limuloides (Bellinurus); 3, Cypris—magnified; 4, Spirorbis (Annelid)—magnified; 5, Phillipsia (Trilobite); 6, Eurypterus (Idothea) Scouleri, from Linlithgowshire.

the open sea; gasteropods, like *euomphalus* and *pleurotomaria*, on shore; acephalans, like *unio* and *anodon*, in freshwater and tidal estuaries; and others, like *aviculopecten*, *mytilus*, and *mactra*, in its shallower bays. The bone-encased fishes of the old red sandstone have now disappeared, and their place is taken by the more fish-like forms (if we may so express it) of *megalichthys*, *palæoniscus*, *amblypterus*, *eurynotus*, and *platysomus;* by gigantic shark-like cestracionts, whose teeth (*helodus*, *pœcilodus*, *psammodus*, &c.) and fin-spines (*gyracanthus*, *ctenacanthus*, *oracanthus*, &c.) are alone preserved to us; and huge sauroid genera (*rhizodus*, &c.), whose dentition marks an affinity to the higher class of reptiles. In many localities these fishes seemed to have swarmed in shoals, preying on shell-fish and

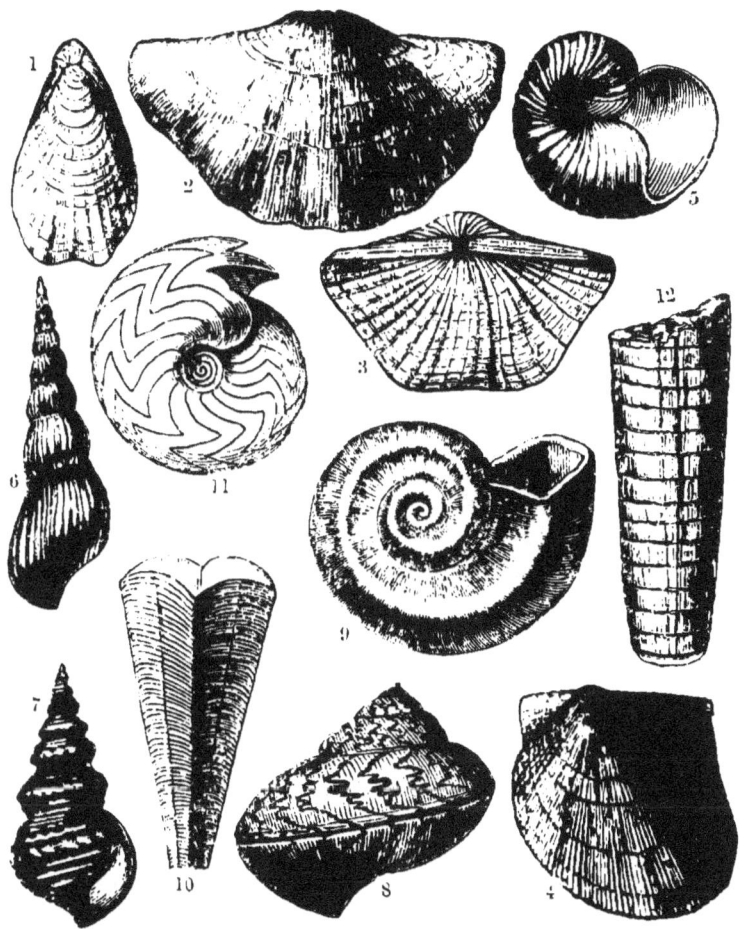

1, Terebratula; 2, Productus; 3, Spirifera; 4, Aviculopecten; 5, Bellerophon; 6, Loxonema; 7, Murchisonia; 8, Pleurotomaria; 9, Euomphalus; 10, Conularia; 11, Goniatites; 12, Orthoceratite.

young coral-growth, and also on one another, as is amply testified by their fossil droppings or *coprolites*, which crowd the shales or muds of the carboniferous sea-bed. In reptilian life, the forms are, on the whole, of lowly organisation, indicating, as it were, the recent advent of the order,—an order whose remains have not been discovered with certainty in any preceding formation. From the European and

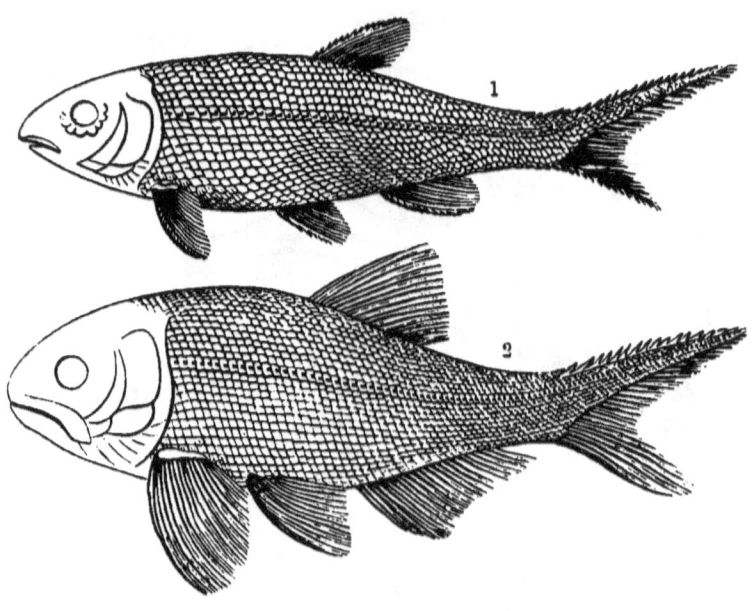

CARBONIFEROUS FISHES.
1, Palæoniscus; 2, Amblypterus.

Nova Scotian coal-fields, however, we have five or six genera of frog-like and lizard-like forms—some evidently aquatic, others amphibious, and some fitted for an arboreal habitat. They are known by such names as *archæogosaurus* (ancient land-lizard), *parabatrachus* (frog-like reptile), and *dendrerpeton* (tree-lizard), and carry the imagination back to stagnant pools, to sludgy river-shores, and to ancient forest-growths, whose hollow trunks furnished at once their insect-food and a place of security and shelter. In these early reptiles—in the persistence of their dorsal chord, their gill-arches, their large median and lateral throat-plates, and other piscine characters—Professor Owen traces a "linking and blending" of the two cold-blooded vertebrate groups; *archæogosaurus* conducting, as it were, the march of life from the fish proper to these labyrinthodont reptiles that come boldly into force in the Permian and triassic eras.

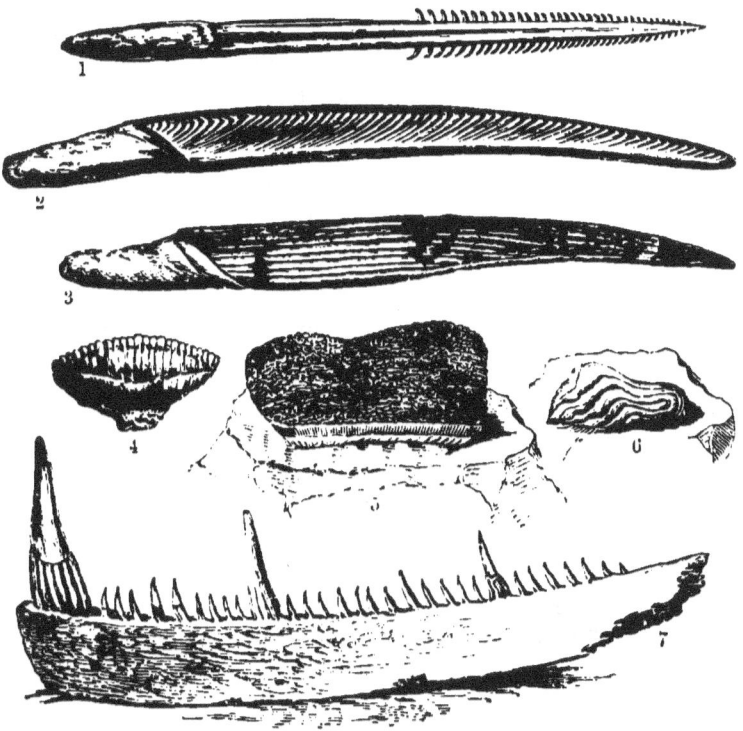

Fin Spines: 1, Pleuracanthus; 2, Gyracanthus; 3, Ctenacanthus. Palatal Teeth: 4, Ctenoptychius; 5, Psammodus; 6, Pœcilodus. 7, Jaw of Rhizodus, showing Reptilian Teeth.

The course of vitality is thus for ever onward and upward —onward in the introduction of forms having more varied geographical adaptations, and upward in the manifestation of higher physiological and functional performance.

Such is the panorama of carboniferous life—an unparalleled exuberance of endogenous flora; a wonderful profusion of estuarine and marine life in all its aspects: but as yet few insects, none of the higher reptiles, no birds, no mammals! And yet, looking at mere external conditions, it is difficult to conceive how, in some of their specific forms, they should not be there. There was abundant food for insects—why not insectivorous reptiles and mammals to prey upon them? Besides insects, there were also fruits and seeds—why not birds to feed upon them; and why not the larger herbivorous reptiles and quadrupeds to browse upon the excess of vegetation that then clothed so large a portion of the earth's surface? True, such plants as equisetums, club-mosses, ferns, and coniferous trees, are, from the peculiar principles they contain, the least fitted for the sustenance of known animals; but then there were the succulent shoots and roots of palms, of calamites, poacites, and other leafy herbage—the fruits of palms and other allied trees, and these we know are the favourite food of many mammals at the present day. Nay more; as we know that certain savage tribes exist on palm fruits, or farinaceous roots, and on the fish of the ocean, we might carry this sort of reasoning still further, and ask whether the human race, in some of its lowlier phases, might not also have been participators in the life of the carboniferous era? To questions such as these the palæontologist has no other answer to offer than that he has hitherto failed to detect the remains of birds and mammals; that as the food to be consumed and the consumer are generally concomitants, so he more than expects the discovery of higher life during the coal-period; but that this higher life, though discovered to-morrow, would necessarily take its stand lower in the scale of organisation than the reptiles, and birds, and mammals which are found in the immediately succeeding formations of the new red sandstone and oolite. If there is one truth that geology has established more clearly than another, it is that of the progressive evolution of life on this globe; not progress from imperfection to perfection, for all are alike fitted to the end for which they were created, but progress from simpler to more specialised forms. All the discoveries that have been made, and are daily making, never controvert in the least this great order of life; nor do the ablest geologists, though anticipating many new forms, ever expect to find it otherwise with creation than onward and still upward. In this respect the coal-formation takes its place orderly and in perfect harmony with what is known of other formations:—more prolific and more specialised in its forms than the old red sandstone beneath, and less so than the new red and other secondary strata that follow.

Looking, in the mean time, at the whole aspects of the carboniferous period, we are reminded (as we have elsewhere* indicated) of geographical conditions never before nor since exhibited on our globe. The frequent alternations of strata, and the great extent of our coal-fields, indicate the existence of vast estuaries and inland seas—of gigantic rivers and periodical inundations; the numerous coal-seams and bituminous shales clearly bespeak conditions of soil, moisture, and warmth favourable to an exuberant vegetation, and point partly to vegetable drift, and partly to submerged forests, to peat-swamps and jungle-growth; the mountain limestone, with its marine remains, reminds us of low islands fringed with encrinite-banks and coral-reefs, and lagoons thronged with shell-fish and fishes; the existence of reptiles and insects tells us of air, and sunlight, and river-banks; the vast geographical extent of the system bears evidence of an equable and continuous climate over a large portion of the earth's surface; while the interstratified trap-tuffs, the basaltic outbursts, and the numerous faults and fissures, testify to a period of intense igneous activity within the same areas, to repeated upheavals of sea-bottom and submergences of dry land. All this is so clearly indicated to the investigator of the carboniferous system, that

* *Advanced Text-Book of Geology.*

he feels as convinced of their occurrence as if he had stood on the river-bank of the period, and seen the muddy current roll down its burden of vegetable drift; threaded the channels of the estuary, gloomy with the gigantic growth of swamp and jungle; or sailed over the shallow waters of its archipelago, studded with reef-fringed volcanic islands, and dipped his oar into the forests of encrinites that waved below.

The Permian period, to which we now turn, presents itself more in the light of a new rock-formation than a distinct life-period. Many of its forms are identical with those of the coal period, and we may, without doing great violence to fact, regard it as the continuation and close of the carboniferous era — specialised by local disturbances in the areas of deposit, and the consequent dying out of many genera and species. Perhaps the most remarkable feature is the rapid disappearance of the coal flora, and its restriction to a few higher forms of tree-ferns and coniferous trees, as if the low swampy jungle had been upheaved into higher and drier lands unfavourable to the growth of

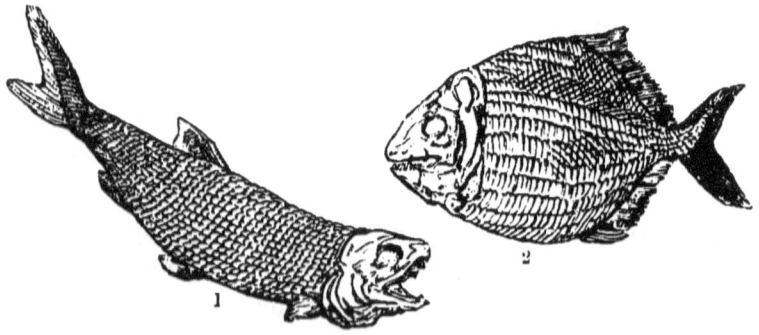

1, Palæoniscus Frieslebeni 2, Platysomus striatus

sigillaria, calamites, equisetums, and lepidodendra. The gigantic sauroid fishes have also disappeared with the estuaries in which they held supreme sway, though less localised forms, as *palæoniscus* and *platysomus*, still occur in abundance; reptiles of larger growth and curious configuration (*labyrinthodon*) come into view; reptilian and bird-like footsteps (*ichnites*) can also be traced on the sandstones; and if American geologists be not mistaken, mammalian life in its lowly marsupial form (*dromatherium*) now comes for the first time on the stage of being. On the whole, however,

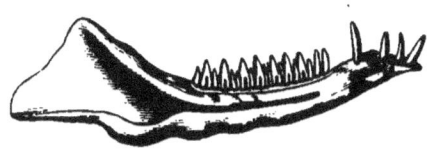

Jaw of Dromatherium silvestre, from the Red Sandstones of North Carolina (Emmons)

there seems a paucity of life during the Permian period, when compared with that which preceded it; and this we may, in the mean time, ascribe partly to geographical changes in the distribution of sea and land, partly to the altered composition of the sea-water in certain areas where we have now magnesian limestones and red ferruginous sandstones, and partly to that change of climate which is indicated by the symptoms of glacial action in the formation of its conglomerates and bouldery breccias.*

* Professor Ramsay, who was the first to advocate, in a decided manner, the glacial origin of these breccias, founds his belief on the following evidences:—1. The great size of many of the fragments—the largest observed weighing (by a rough estimate) from a half to three-quarters of a ton. 2. Their forms. Rounded pebbles are exceedingly rare. They are angular or sub-angular, and have those flattened sides so peculiarly characteristic of many glacier-fragments in existing moraines, and also of many of the stones of the pleistocene drifts, and the moraine matter of the Welsh, Highland, Irish, and Vosges glaciers. 3. Many of them are highly polished, and others are grooved and finely striated, like the stones of existing Alpine glaciers, and like those of the ancient

We now close the long record of Ancient Life, during which whole races and families departed, and others took their place—the march of vitality being ever forward to higher and higher orders. We have seen that all the great types of life — radiate, molluscan, articulate, and vertebrate—had their beginning simultaneously and independently on the globe, and that all subsequent progress has been restricted to the modification and elimination of these primal patterns. We have seen the *graptolites* of Siluria rise, culminate, and depart with that period; seen also its curious *encrinites* and foot-stalked sea-urchins, or *cystidew*, flourish and die within the same limits; and witnessed its wonderful flush of *trilobite* life, which waned in the old red, and finally disappeared about the middle of the carboniferous era. So also have we witnessed the larger crustacean forms of *eurypterus* and *pterygotus* come strongly and forcibly on the Devonian stage, and somewhat speedily wane and die out with the coal period, during which other forms, prefigurative as it were of the existing *limulus*, take their places. In like manner the curious bone-clad fishes of the old red (the "palichthyan" aspect of fish life) rise and depart with that system—only a few of the genera, but none of the species, living into the carboniferous epoch. And when we come to the coal period itself, there also all the wonderful and exuberant forms of its vegetation — its *stigmaria, sigillaria, lepidodendra, bothrodendra, calamites,* and *tree-ferns*— start into being, flourish in profusion, and depart with those physical peculiarities which stamped

glaciers of the Vosges, Wales, Ireland, and the Highlands of Scotland; or like many stones in the pleistocene drifts. 4. A hardened cementing mass of red marl, in which the stones are very thickly scattered, and which in some respects may be compared to a red boulder-clay, in so far that both contain angular, flat-sided, and striated stones, such as form the breccias wherever they occur.—*Journal of Geological Society*, vol. xi.

their impress on the life of that era. So also with its sauroid fishes; and so also with many genera and species of its shell-fish and corals and encrinites, which though more lowly are nevertheless peculiarly distinctive of carboniferous seas, and are never found in the waters of subsequent ages.

From the first to the last—from the Silurian to the Permian—all has been growth and decay, and in that death a progress which ever goes forward without halt or hesitation. No indecision; no trial-work; no error to be corrected; no blunder to be revised. And yet amid all this incoming and outgoing, as we shall see in the following chapter, there has been no break in vitality, no change of the great primal patterns, but merely such modifications as best harmonise with the new conditions of each succeeding era. Nor must we regard this harmony between geographical condition and organic manifestation in any other light than that of a mere co-adaptation; for over and above it there is clearly a prescient design, having respect to development in time from more general to more specialised types, and from physiological simplicity to physiological complexity of functions. From the obscure and simple forms of the lowest stratified systems we rise stage after stage to higher and higher manifestations of life; onwards and still upwards is the orderly course of creation; and yet in this vast and varied progression every member is bound to that which preceded it, as well as to that which accompanies it, by the ties and relationship of one great cosmical plan. This is surely more than mere " physical development"—something higher than the " transmutation of specific forms under the force of external conditions"—something more precise and definite than " natural selection in the struggle for existence," or any other of the materialistic hypotheses that have been recently advanced to account for the great chronological elimination of vitality. It is (if anything man shall ever comprehend) the gradual unfolding of a predestined scheme—a divine conception, to the realisation of which the various forces of nature, co-related and co-adapted, are in ever-active co-operation.

THE MIDDLE PAST.

MESOZOIC SYSTEMS—THE TRIAS, OOLITE, AND CHALK.

WE now take leave of the palæozoic aspects of the world, and pass on to those of the Mesozoic or "middle life" period—characterised by forms and species which hold an intermediate place between those of the more ancient and those of the more modern epochs. The grand primeval types and patterns are still the same—radiate, molluscan, articulate, and vertebrate—but the modifications of the types are new, and the consequent organisation higher and more complex. The "differentiation" of the vital functions (as zoologists express it) now becomes more marked and apparent—that is, instead of organisations in which several functions are performed by the same organ, each function has an organ specially devoted to its purpose. The expression of Creative thought has become more specialised, and the plants and animals of the newer epochs bear the impress of that specialisation, and find in new external conditions a fitting habitat for their growth and elimination.

We now take farewell of the graptolites, cystideans, trilobites, and eurypterites of Silurian seas—of the gigantic crustaceans and bone-cased fishes of the old red sandstone —of the sigillariæ, stigmariæ, lepidodendra, and other endogenous forms of the coal period—of the cup-in-cup, honeycomb, chain-pore, spider-web, and other corals of the Devonian and mountain limestones—of the huge reptile-like fishes that swarmed in carboniferous waters; and are introduced to other species and newer forms of vitality. The vegetation that adorns the lands of the mesozoic period bears a closer resemblance and affinity to the tree-ferns, cycads, zamias, palms, and subtropical pines of the present day; and the botanist feels he can now institute comparisons with some prospect of success, and attempt restorations with greater confidence and certainty. So also in the animal world the approximations are becoming closer and closer; the divergence from existing families is less perceptible even to the unscientific observer; and the zoologist now meets with all the great divisions of vertebrate life—fishes, reptiles, birds, and mammals. A vast progress has been made in the great onward evolution of vitality—whole families of lower life have died out, and higher ones have taken their places—and orders only beginning to come into existence in the primeval world are now approaching their culmination, or point of greatest numbers, variety, and development.

Besides these gradational advances from lower to higher forms, which are common to every geological epoch, there are also some curious external characteristics which must arrest the notice even of the least scientific and the least geological of observers. So noticeable are these features, that if the fossils of the palæozoic cycle were arranged on one side of a museum, and those of the neozoic on another, the difference would strike the casual observer as strongly as would the difference between the brute-man sculptures of the Ninevites and Egyptians on the one hand, and the man-god sculptures of the Greeks and Romans on the other. It is like passing from the Assyrian and Egyptian chambers of the British Museum to those devoted to the Greeks and

Romans. The expression of human thought is not more clearly indicated by the remains of these ancient civilisations, than the expression of creative thought is indicated by the fossil forms of the palæozoic and mesozoic Earth-periods. Thus, in the palæozoic endogens the ultimate development of the leaf is, for the most part, stamped in permanent beauty on the tall sculptured stems, whereas in the neozoic exogens it ascends to the more exquisite but evanescent beauties of the flower and fruit. Again, the palæozoic leaf, being endogenous, has a venation wholly parallel, whereas the neozoic leaf adds the reticulated venation of the exogen to that of the endogen. Further, as the floral arrangement of the endogen is formed by *three*, and that of the exogen by *five*, all the palæozoic flowers and fruits are stamped by the normal number *three*, whereas *fives* and *threes* are equally normal in the neozoic flora. So also in the animal kingdom : the corals of the palæozoic cycle had their septa or ray-like partitions arranged in *fours*, while those of the neozoic are arranged in *sixes*. In the palæozoic cephalopods the arms are for the most part void of sucking discs, while those of the neozoic seas are, on the other hand, generally furnished with them ; and in the chambered shells of the same order, the palæozoic species have their sutural junctions plain and simple, while those of the neozoic are often foliated and of most intricate pattern. The palæozoic crustacea, even in the highest forms yet discovered, are more larval-like or abdominal in their segmentation than the neozoic, in which head, thorax, and abdomen become distinct and definite. Again, the palæozoic fishes had all the heterocercal or unequally-lobed tail (which marks the embryonic condition of fish-life in general), while in the neozoic order, the heterocerque is subordinated, and the homocerque, or equally-lobed, and the undivided tails become the general and normal forms.

These and other distinctions, upon which our limits will not permit us to dwell, stamp the palæozoic as a life-period widely different from that of the mesozoic, and yet there was no break, no discontinuity in the great evolution of vitality. As the life of one system runs imperceptibly into that of another, and the two have always some forms in common; so the palæozoic runs into the mesozoic, and it is only when viewed as a whole, and at a sufficient distance, that its distinctive characters stand out in bold and peculiar relief. So in like manner we shall find it with the mesozoic life-period, when we have reviewed the forms of its triassic, oolitic, and cretaceous systems. It has a facies peculiar to itself, and though approaching in some of its features, yet as a whole unmistakably different from the facies of the cainozoic period, which is now running its course, and bearing us along with it.

And first, we turn to the Trias or upper new red sandstone, with its "triple" series of various coloured sandstones, shelly limestones, and saliferous and gypseous shales. These party-coloured deposits, in which ferruginous tints predominate, are clearly the sediments of circumscribed oceanic areas—areas which, in the northern hemisphere at least, were of no great depth, and subjected to repeated elevatory and depressing movements. This new arrangement of sea and land, accompanied by no gigantic rivers or estuaries, and apparently by a somewhat arid climate, is characterised by a numerical as well as specific paucity of life—a paucity which is greatly aggravated by the unsuitable nature of the sandstones and marls for the preservation of organic remains. Physiologically, however, the forms are still on the advance; cycads and conifers are more decided in their characters; brachiopods diminish, and true bivalves increase; cold-blooded air-breathers become more numerous, and warm-blooded races (birds and marsupials) for the first time make their unmistakable appearance. Vitality, in obedience to some great law of progress, is ever pressing forward to higher and higher forms, even though restricted to unstable seas, and subjected to the stunting influences of riverless plains and thirsty uplands.

Of the marine flora of the trias we know little or nothing. *Fucus*-like impressions are occasionally retained on the sandstones, but so fragmentary and obscure that a "general re-

FRAGMENTS OF TRIASSIC FLORA.
1, Walchia ; 2, Pterozamites or Pterophyllum from North America (Emmons)

semblance" is all the palæontologist can affirm. When we turn to the land-plants, equisetums, calamites, ferns, cycads, and conifers are the predominating forms;—the equisetum and calamite pointing to the marshy pools of the summer-dried river-course, the fern and cycad to the scrubby plain, and the coniferous trees to the open upland. The triassic *equisetums, calamites*, and tree-ferns (*sphenopteris*, *pecopteris, neuropteris,* &c.), though bearing the stamp of generic resemblance to palæozoic forms, and evidently fulfilling the same functions, are all of different species; while the cycadaceous *pterophyllum* and *Mantellia*, and the coniferous *Voltzia* and *Walchia* are altogether new and unknown to former floras. On the whole, the aspect of the triassic flora is more akin to that of the oolite, which succeeds, than to that of the carboniferous that went before; and though scantily exhibited in the areas of Britain and Germany, many have had a fuller and more connected development in other regions. At all events, we are not entitled to generalise from these limited localities, but rather to believe that the apparent severance between palæozoic and mesozoic was bridged over by intermediate forms that now lie entombed in areas still unknown or covered by the existing ocean.

In the animal kingdom (the forms being chiefly marine) the connection is more continuous and intelligible, even though the bulk of triassic sediments are highly unfavourable to the preservation of organic structure. The lower Radiate forms are yet little known, few corals occurring in any investigated area, and only two or three species of *encrinite* and *pentacrinite*. The higher radiates are equally rare, there being no well-authenticated instance of a sea-urchin, and only two or three genera of star-fish, as *ophiura, aspidura,* and *asterias*. The Articulata are even still more scantily represented in the triassic seas of Europe, only a few insignificant *serpulæ* and a single crustacean (*palinurus*) being all that have yet turned up to the palæontologist; thus leaving an almost unbridged gulf between the higher annelids, crustacea, and insects of the coal, and those that are known to succeed in the oolite. This, however, is obviously a local imperfection in the Record, and geologists look forward with interest to the discovery of the connecting forms in the strata of other triassic regions. When we turn to the Mollusca, the record becomes much more satisfactory and connective, though many of the old genera are evidently on the wane, and several have wholly departed. The brachiopods, diminishing alike in generic and numerical force, are still represented by *lingula, terebratula,* and *spirifer;* the conchifera, or true bivalves, are vastly on the increase, and such forms as *trigonia, mya, plagiostoma, avicula,* and *ostrea,* throng the waters; the gasteropods present *buccinum, turbo, turitella,* and other characteristic genera; while the predaceous cephalopods, rising in complexity of structure, are represented by *orthoceras, nautilus, ceratite, belemnite,* and *rhyncholite.* The Fishes of the pe-

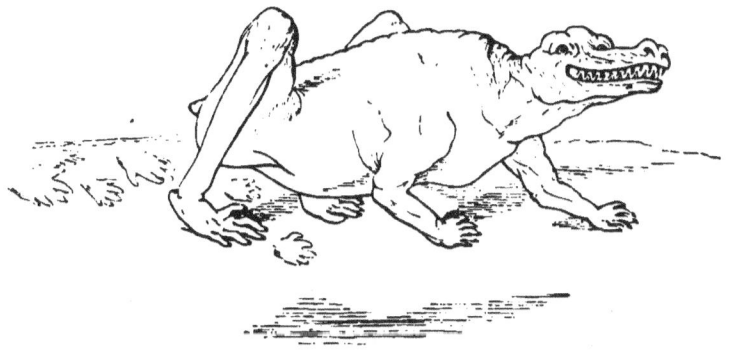

Restored form of Labyrinthodon, with footprints the same as Cheirotherium.

riod present as yet few well-determined forms, being known chiefly by their detached teeth and fin-spines. These organisms, scattered as they are, clearly point to shark-like genera (*ceratodus, hybodus,* &c.), whose mouths, like that of the Australian cestracion, were paved with broad-crowned corrugated teeth for the crushing of shell-fish, while their serrated fin-spines supplied them with a sure and ready means of defence. Besides these, there are other forms (*saurichthys,* &c.) which still carry forward, though on a diminished scale, the line of sauroid fishes that had its culmination in the estuaries of the carboniferous era. But if sauroid fishes are on the wane, true reptiles—marine and amphibious—are strikingly on the increase, their teeth, bones, and footprints foreshadowing that enormous development and variety that found its meridian during the oolite epoch. Of these the gigantic frog-like *labyrinthodon*, the *plesiosaur*, *phytosaur*, and *thecodontosaur*, the small lizard-like *telerpeton* of the Elgin sandstones, the larger *hyperodape-*

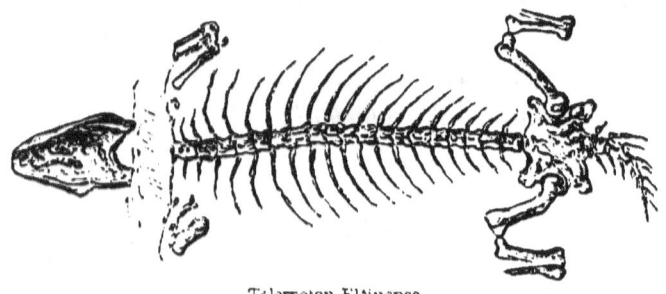

Telerpeton Elginense

don, and the crocodile-like *stagunolepis* of the same formation, are perhaps the most noticeable forms; while innumerable foot-tracks (*chelichnus, cheirotherium, batrichnis*, &c.) point partly to turtle-like, partly to frog-like, and partly to crocodilian-like genera.

Still higher in the scale of life rank the foot-tracks of gigantic birds, and the teeth and jaws of small insectivorous mammals. These fossil foot-prints (*ichnites*) form one of the most curious features of the period, and their study (*ichnology*) constitutes one of the most interesting chapters in geology.* In the successive stages of the earth's history

* The valley of the Connecticut in America, Corncockle Muir in Dumfriesshire, Storeton in Cheshire, and Hildberghausen in Germany, have been, as yet, the chief repositories of these fossil footprints. So abundant are they in the Connecticut sandstones, which are mainly triassic (the upper being of the age of the lias, and the lower perhaps permian), that

worms must have tracked and burrowed in the open sands, shell-fish and crustacea crawled and pattered on the muddy beach, and reptiles, birds, and mammals footed the tidal silt of bays and estuaries. Wherever these materials were of sufficient consistence, and exposed during a long tidal ebb to the desiccating effects of the sun, there the impressions would be retained, and act as a mould for the reception of the next influx of mud. The mould and its cast, covered over by repeated sediments, are thus preserved for ever, bearing every outline of form and minutiæ of structural surface, according to the nature of the deposit that received the living impression. Over these old triassic shores numerous birds and reptiles waded and wandered, now wheeling in sport, now fleeing in fear, and anon stealing stealthily on their devoted victims. Not a bird-bone has yet been discovered in the sediments that bear these fossil footprints,* and yet so characteristic is the foot of the bird (the number and disposition of its joints, and the corrugations of its skin), that palæontology rests satisfied in their existence as fully as though their skeletons were there to indicate their habits and dimensions. In the case of mammalian life the evidence, though scantier, is much

Dr Hitchcock has already enumerated 123 species—viz., marsupialoid animals, 5 ; birds, 31 ; ornithoid reptiles, or reptiles walking on their hind feet, 12 ; lizards, 17 ; batrachians or frog-like reptiles, 16 ; chelonians or turtles, 8 ; fishes, 4 ; crustaceans, myriapods, and insects, 17 ; and annelids, 10.

* In 1860, a block of red sandstone, containing the impressions of bones, apparently of ornithic character, was discovered in America, and described by Professor W. B. Rogers to the Natural History Society of Boston. This block was not, however, found *in situ*, but among other building stones which were said to have been brought from Portland Quarry, in the valley of the Connecticut. The evidence is thus somewhat invalidated, though Professor Rogers seems confident as to its mesozoic or new red sandstone origin. This specimen, unique in the mean time, gives hope of the speedy discovery of other and more legible fragments.

more conclusive—teeth, jaws, and other fragments pointing unmistakably to small marsupial quadrupeds (*microlestes, dromatherium*, &c.), which find their nearest analogues in the wombats and kangaroo-rats of Australia.

Such is the scanty and imperfect record of triassic life, as preserved in the variegated sandstones, the muschelkalk, and saliferous marls of Europe and North America. This imperfection may arise, partly from the circumscribed and varying seas of deposit, partly from the saline peculiarities of their waters, and partly from the unsuitable nature of their sediments for the preservation of organic structure; but from whatever cause, we are clearly not entitled to generalise from these limited areas to the universal distribution of triassic vitality. On the contrary, the steady creational advance to higher and higher facies of life, presuppose not only an extensive series of gradational species, but a numerical exuberance through which the law of specific advancement could operate. And even now, in the St Cassian beds of the Austrian Alps, we are not without evidence of many new and connecting forms—genera which unite the palæozoic and mesozoic into one continuous lifestream, and forbid the unphilosophical idea of creational breaks in the evolution of vitality. As the facts stand (and we know little beyond a few unconnected belts of deposit in the northern hemisphere), the triassic flora points to insular rather than to continental conditions, and to an arid rather than to a genial climate; its marine fauna, in harmony with these conditions, points rather to circumscribed seas than to gigantic estuaries; whilst its terrestrial animals indicate the thirsty desert rather than the fertile plain, and sun-baked muddy creeks rather than the exposed shores of the open ocean.

The Oolitic era, to which we next turn, presents geographical conditions extremely different from those of the trias, and apparently more favourable to an exuberant exhibition of vitality. In its ascending series of Lias, Oolite, and Wealden, we have a succession of deep-sea, littoral, and estuarine deposits, which, in the old world at least, spread over wider areas, and are in their calcareous muds, clays, and limestones, much more conservative of organic structure. Not only do the seas show broader and more southerly expanses, but they are more connected, and seem to have been less liable to sudden variation either in their depth or configuration. Their waters were likewise more normal in their composition, and we get quit of those super-saline and ferruginous constituents which, in the trias, appear to have been as unfavourable to the development as to the preservation of organic nature. The land also assumes a more continental aspect, and, under a genial climate, gigantic rivers and estuaries bespeak conditions conducive alike to numerical abundance and specific variety. In conjunction with these new conditions of area and climate, vitality puts on newer aspects. The palæozoic forms that lingered in the trias altogether disappear, and mesozoic life, in a prolific flora and fauna, attains its meridian of development. In vegetation, palms, lilies, and other allied monocotyledons are on the increase; tree-ferns, cycads, and conifers, are the dominant orders; while dicotyledonous types make their appearance in fragments of wood, leaves, and inflorescence. In the animal kingdom the advance is still more marked and decisive. Zoophytes and other lowly orders are more abundantly and beautifully preserved; sea-urchins, starfish, and crustaceans assume generic aspects more akin to existing races; bivalves and gasteropods are still largely on the increase, and cephalopods attain their specific and numerical meridian; the fishes more closely approximate the existing ichthyic type; and though indications of mammalian life become more abundant, reptiles—aquatic, terrestrial, and aërial—herbivorous, carnivorous, and omnivorous—are now the dominant forms, and discharge in their every function the part now assigned to the several grades of the higher mammalia.

The marine plants of the oolite, like the marine flora of all other geological formations, are indistinct and fragmentary. Their bifurcating impressions are not unfrequent in some of the oolitic sandstones, but such names as *halymenites* indicate a resemblance rather than a determinable affinity to any living form. Aquatic plants, resembling the pond-weeds (*chara, naiadites,* and the like), occur in considerable abundance, but little has been done to fix their true relations to existing orders, and in the mean time we can do little more than note the fact of their presence, and indicate the conditions that must have favoured their development. Among the lower or cryptogamic orders of land-plants, equisetums (*equisetites*), and club-mosses (*lycopodites*), though not so frequent as in earlier formations, are by no means uncommon; while tree-ferns (*pecopteris, sphenopteris, tæniopteris, otopteris,* &c.) appear in vast profusion, and many of them peculiar to and restricted to the period. Stems and leaves of unknown endogens (*endogenites*), palms (*palmacites*), and lily-like plants occur throughout the formation, while cycadaceous stems, leaves, and fruits (*cycadeoidea, palæozamia, zamites, pterophyllum, zamiostrobus,* &c.) constitute one of the most noticeable botanical peculiarities of the period. Coniferous trees are also in the ascendant, and so similar in many respects to the cypresses, araucarias, thujas, yews, and pines of southern latitudes, that their affinities are at once expressed by such terms as *cupressites, araucarites, thuyites, taxites,* and *pinites*. Altogether, the vegetation of the oolite presents a high specific as well

OOLITIC ERA. 131

as numerical abundance, and indicates genial and continuous geographical conditions—so genial as to give rise in

RESTORED ASPECT OF OOLITIC VEGETATION.
Palm, Screw-pine, Araucaria, Cycas, Tree-fern, &c.

many areas (Europe, India, the Indian Islands, and North America) to repeated and valuable deposits of coal. Indeed, many coal-fields at one time attributed to the carboniferous epoch have been proved to be of oolitic age,* and, as investigation is pushed still further, other areas, in both hemispheres, will be found to belong to the same geological system.

When we direct our attention to the fauna we find the lower marine animals abundantly represented, showing that in the oolitic seas there were those varied conditions of warmth, depth, sea-bottom, and shore-line essential to their dissemination and development. Sponges (*spongia*) are by no means rare; foraminiferous organisms (*lituola, rotalina, spirolina*, &c.) are scattered throughout the formation; and corals (*thamnastræa, montlivaltia, isastræa*, &c.) of varied and elegant forms occur in vast profusion, and point to a time when the oolitic areas of Europe and Asia were instinct with coral-life, and dotted and barred with reefs like the existing seas of the southern hemisphere. Encrinites, though now on the wane, still star the sea-bed with their elegant forms (*pentacrinus, apiocrinus*, &c.); sea-urchins (*cidaris, hemicidaris, diadema, echinus*, &c.) throng the marine strata in increasing numbers; and free-floating star-fishes (*astropecten, amphiura*, and *ophioderma*), apparently replacing the encrinites, now approximate in generic aspects to those of the present ocean. Annelids, like the living *serpulæ*, cement their tortuous tubes to stones and dead-shells; barnacles (*pollicipes*) attach their many-valved mansions to rocks and floating timber; minute crustaceans (*cypris, cypridea*, and *estheria*) moult their bivalved crusts in myriads in the muddy creeks and estuaries; while the higher crustacea (*glyphæa, eryon*, and *megacheirus*) approxi-

---

* The coals of Southern India, of Borneo, Labuan, Zebu in the Philippine Islands, &c., are now ascertained to be of oolitic age; to which epoch also it is suspected that most of those of China and Japan belong; as well as that of Virginia in America, and other localities. The oolitic coal-fields of Eastern Yorkshire and Brora in Sutherlandshire have been long known to British geologists.

mate in form and function to the crayfish and lobsters of existing waters. Insects, destructible as their remains may

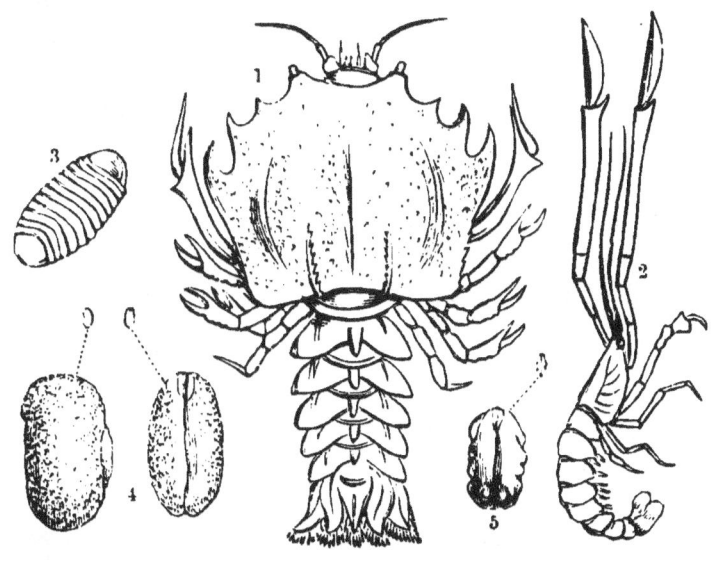

OOLITIC CRUSTACEA.
1. Eryon; 2. Megacheirus; 3. Archæoniscus; 4, 5, Cyprides—natural size, and magnified.

seem, now assume an important place in the lists of the palæontologist—burrowers among the decaying timber of the pine-forests; leapers among the leaves and herbage of the cycas grove; hunters along the river-bank and across its sunny waters; and gaudy flutterers over the flowers of the lily and palm-tree. All the great orders of insect-life —beetles, cockroaches, dragon-flies, grasshoppers, and ants —are abundantly represented, and their resemblances (if not affinities) are indicated at once by such generic appellations as *buprestium, blattidium, libellelium, cicadellium,* and *formicium.*

The waters are now thronged with molluscan life. The minute polyzoans or sea-mats weave their delicate network (*diastopora, ceriopora, heteropora,* &c.) over shells, encrinites, and every available ground-work—varying slightly in

I

pattern, but still preserving that similarity of design which has ever characterised their beautiful structures. The deep-sea, infusorial-feeding brachiopods, though specifically fewer than in the palæozoic periods, are still abundantly represented—*terebratula, rhynchonella, spirifera, discina*, and the like, being the dominant forms in the marine beds of

OOLITIC MOLLUSCA.
1, Spirifer; 2, Avicula; 3, Terebratula; 4, Pholadomya; 5, Modiola; 6, Gryphæa; 7, Trigonia; 8, Plagiostoma; 9, Pleurotomaria.

the lias and oolite. The true bivalves, now so greatly in the ascendant, present themselves in vast profusion, thronging every condition of sea-shore, and leaving their remains in every degree of beauty and perfection. *Gryphæa, gervillia, avicula, lima, ostrea,* and *pecten; trigonia, modiola, pholadomya, cardium, astarte,* and scores of other genera, occur in numerous specific forms in the marine beds of the lias and oolite; while fresh-water mussels (*unionidæ*) are equally characteristic of the estuarine sediments of the

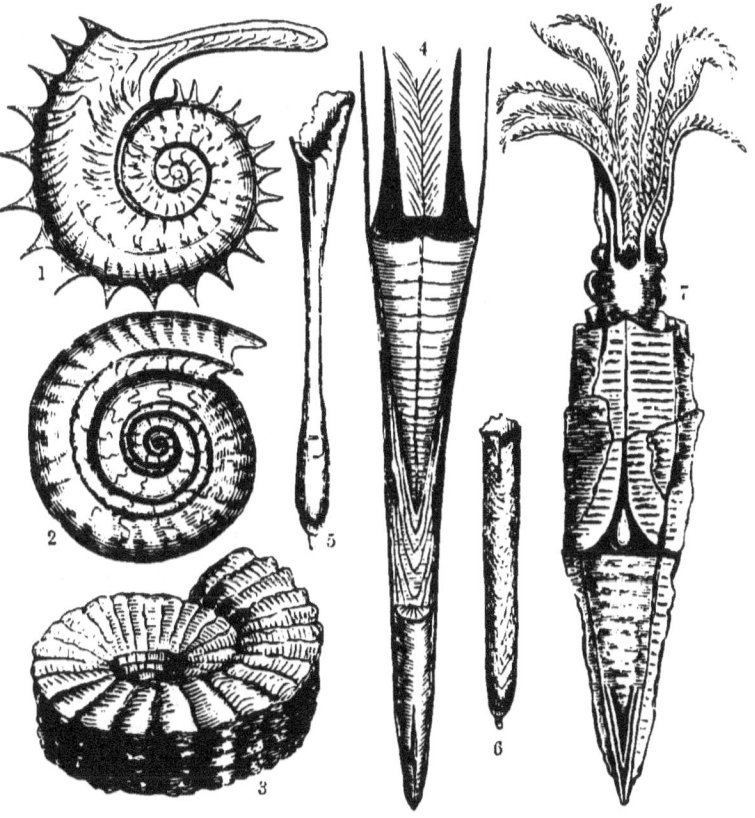

OOLITIC CEPHALOPODS.
1, Ammonites Jason ; 2, A. communis ; 3, A. Bucklandi ; 4, Belemnites Puzosianus 5, 6, Belemnites ; 7, belemnoteuthis.

Wealden. The gasteropods, too, in many generic aspects, crowd the sea-shores—*turbo, trochus, pleurotomaria, nerinæa, patella, cerithium*, and the like, being the more common forms in the marine strata; while *planorbis, paludina*, and their congeners occur in the fresh-water limestones of the Weald thick as their modern species do on the marls of our lakes and marshes. The cephalopods now attain their meridian, and in variety of form, size, and numbers, stamp the period with one of its most peculiar aspects. Shell-clad genera, like *nautilus* and *ammonite*, leave their chambered habitations in myriads; and naked genera, like the cuttle-fishes, are evidenced by thousands of those internal organisms (*belemnites*) which survive the decay of the softer structures. It is indeed the "reign of ammonites"—these beautiful shells occurring in hundreds of specific forms, in every stage of growth, and in the most diversified styles of external ornamentation. Along the exposed shore, in the land-locked bay, and out in the open waters of the old oolitic seas, these predaceous shell-clad cephalopods reign the lords of molluscan life, and mark the culmination of an order which now finds its only representative in the plain-looking nautilus of the Southern Ocean.

The fishes to which we next ascend belong exclusively to the great placoid and ganoid divisions—the soft-scaled orders (the ctenoids and cycloids) of the newer epoch being as yet unrepresented in oolitic waters. The placoids are chiefly rays and sharks, whose teeth (*hybodus, acrodus, ganodus*, &c.) and fin-spines (*asteracanthus, nemacanthus,* and *myriacanthus*) were the only preservable portions of their uncalcified skeletons. Many of these, like the cestracion of the Australian seas, were evidently fitted for the crushing of crustaceous and testaceous animals, others for the prehension of fishes, while some, more slenderly armed, gorged themselves, like their modern congeners, on the squids and cuttle-fishes that then thronged the ocean. The ganoids, now more ichthyic in their aspect, appear in numerous generic forms (*pachycormus, pycnodus, echmodus, lepidotus*, &c.), the majority of which are peculiar to the mesozoic period, and many of them even restricted to the time of the lias and oolite. It is now, too, the high noon of reptilian development—" the Age of Reptiles"—when marine genera (*ichthyosaurus* and *plesiosaurus*) were the

Ichthyosaurus; Long-necked Plesiosaurus; and Pterodactyle.

whale-like monarchs of the ocean; when crocodilians (*teleosaurus* and *cetiosaurus*) thronged the rivers and estuaries; turtles (*chelone* and *platemys*) traversed the muddy shores; gigantic land-saurians (*megalosaurus, hylæosaurus,* and *iguanodon*) roamed, elephant-like, over the river-plains, or browsed in the virgin forest; lizards (*lacerta* and *macellodus*) basked on the sunny cliffs; and bird-like genera (*pterodacty-*

OOLITIC LAND REPTILES.
Restored forms of Megalosaurus and Hylæosaurus—HAWKINS.

lus) winged the upper firmament. Every adaptation of form and function finds its exemplar in these ancient saurians, and the part now played by birds and mammals was then in a great measure discharged by reptiles. They were the representatives in time of the higher orders of vitality— occupying every habitat, aquatic, terrestrial, and aërial, and fulfilling every function, herbivorous, carnivorous, and omnivorous. Everywhere they are the dominant forms, and though birds and mammals are coming more clearly on the stage, the great vital phase of creation was, for the time being, unmistakably reptilian.

This "Reign of Reptiles," as it is sometimes termed, has suggested to minds, more imaginative than logical, the idea of an epoch of incessant warfare and murder; and nothing is more common than pictorial delineations and high-wrought descriptions of reptilian carnage and cruelty.

Transferring the attributes of the infuriated human mind to the unreasoning brute, they picture every species lying in wait for his neighbour—writhing in savage combat for supremacy, and mangling with their horrid fangs even where prey does not become a necessity. Alas! for man's mistaken notion of creation's life-scheme; as if, even in a world of reptiles, there were not a thousand checks and compensations ever actively at work to secure the greatest happiness of the greatest numbers. No doubt the flesh-eater preyed on the plant-eater, and the weak succumbed where the strong exulted; but death comes unconsciously quick where the preyer strikes from necessity, and the fall of the sickly gives wider verge to the enjoyment of the healthy survivor. The wants of nature supplied, and then, as now, the gigantic herbivora rolled sportively among the over-topping herbage, or stood drowsily dreaming under the shade of the noonday forest; while the carnivora gambolled in the open waters, or lazily sunned themselves on the ebbing sea-shore. Wherever life prevails, there also is meted out to it its measure of enjoyment, and man only errs when, describing the lower animals, he invests them with passions and feelings unfortunately too frequently his own. But cold-blooded air-breathers, however varied in size, form, and function, were not destined to be the culminating orders in the world's life-scheme. The divine creational idea, fixed from the beginning, was steadily evolving itself into higher and higher types; and along with this overwhelming exuberance of reptiles, the line of triassic birds was continued in such forms as *palæornis* (ancient-bird), while in certain areas there appeared the higher manifestations of mammalian development. Small insectivorous quadrupeds—*amphitherium* (doubtful-beast), *phascolotherium* (pouched-beast), *stereognathus* (thick-jaw), *plagiaulax* (oblique-grooved tooth), &c.—have been detected in the upper oolite, apparently marsupial in their structure, and pointing to the wombats, bandicoots, and phalangers of Australia as their nearest living analogues. From the number of these imbedded in a few square yards of a stratum near Swanage in Dorsetshire, we may confidently look forward to the discovery of many other mammalian forms—every condition of the period being favourable to the development of such a fauna.

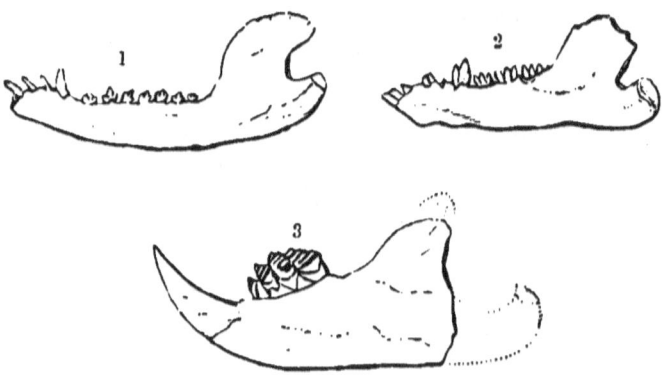

Oolitic Mammals, natural size—1, Lower Jaw and Teeth of Phascolotherium;
2, Of Triconodon; 3, Of Plagiaulax.

Such are the phases of oolitic life, and such the conditions of sea and land, which its miscellaneous sediments seem to imply. Continuous lands of ample area for the growth of a varied flora, open free-flowing seas for an exuberant marine fauna, gigantic estuaries and river plains for the amphibious reptiles of the Weald, and over all a genial but periodically interrupted climate. We have as yet no means of determining the universal climatology of the period, but over the oolitic areas of the northern hemisphere the varying rings of coniferous growth would seem to indicate seasonal variations, while the prevailing aspect of the flora, the abundance of land reptiles, and the presence of small marsupials, point to conditions of general warmth and periodic drought, such as now obtain over the riverless plains of

Australia. As in its external, so in its vital conditions the oolitic epoch finds its newest analogue in the flora and fauna of the Australasian continent, thus indicating once more the connection that invariably subsists between the manifestations of life, and the conditions by which they are surrounded. "The close approximation," remarks Professor Owen, "of the amphitherium and phascolotherium to marsupial genera, now confined to New South Wales and Van Dieman's Land, leads us to reflect upon the interesting correspondence between other organic remains of the British oolite and other existing forms now confined to the Australian continent and adjoining seas. Here, for example, swims the *cestracion* which has given the key to the nature of the palates from our oolite, now recognised as the *teeth* of congeneric gigantic forms of cartilaginous fishes. Not only *trigoniæ*, but living *terebratulæ* exist, and the latter abundantly, in Australian seas, yielding food to the cestracion, as their extinct analogues doubtless did to the allied cartilaginous fishes called *acrodi*, *psammodi*, &c. Araucariæ and cycadaceous plants likewise flourish on the Australian continent, where marsupial quadrupeds abound, and thus appear to complete a picture of an ancient condition of the earth's surface, which has been superseded in our hemisphere by other strata, and a higher type of mammalian organisation." This picture, however, must be received as nothing more than the merest analogy. Nature never repeats herself in time any more than in space, and forms once gone disappear for ever. To speak, as some have done, of Australia being "a belated portion of the earth's surface," is altogether to misinterpret the scheme of creational progress. The species of the oolite are not the species of Australia, while fossil evidence already shows that the present races of the Austral islands have had their gigantic tertiary predecessors, just as other regions have had theirs, and this in a genetic line backwards through the prior epochs of the chalk and oolite. In some of its minor features the oolite may find an analogue in existing nature, but in its entirety it stands alone—a great life-epoch, whose forms are not to be confounded either with what has gone before, or with what has yet to follow.

The Cretaceous or Chalk period, to which we next turn, brings to a close the long and exuberant line of mesozoic life. Great changes in the relative distribution of sea and land in the northern hemisphere have been gradually brought about; much of the oolitic sea-bed has become dry land; and the areas of deposit have assumed a less southerly aspect. Stretching more in an easterly and westerly direction, they present less variety of climate, and, opening up to the north, they become recipients of currents which tend to deteriorate the more genial conditions of the oolitic era. Greensands, clays, clay-marls, and chalk of varying consistence form the prevailing sediments, which, being eminently marine, are replete with the remains of oceanic life. Little of the terrestrial surface of the period is indicated by the fossil flora or fauna, and much of the marine area in Asia and in America has been but imperfectly explored. Notwithstanding this imperfection of the record, we find enough to corroborate the ever-onward progression of vitality, and to show that oolitic forms, though by no means rare, are gradually being replaced by others peculiar to the chalk and greensand.

The Flora, though scantily preserved, has still somewhat of an oolitic aspect, looking more like the remnants of that age than the peculiar products of a newer epoch. Sea-weeds (*confervites* and *chondrites*) resembling the living confervæ and Irish-moss, ferns (*lonchopteris*), lily-like leaves (*dracæna*), cycads (*zamiostrobus* and *clathraria*), and coniferous trees (*abietites* and *strobolites*), in drifted fragments, are all or nearly all we can read intelligibly of the cretaceous vegetation. And yet we know, from certain lignitic beds, that considerable areas must have been clad with swamp and forest-growth, and this during long periods of alternate reproduction and decay.* On the whole, however, the cretaceous flora appears to have been by no means an exuberant one—less varied in its form than that of the oolite which preceded, and less southern in its aspect than that of the early tertiary that followed. And the cause of this we find in the colder currents of the northward opening seas—seas which occasionally brought drifting-ice even to the latitudes of the British Islands—if we are to seek in ice-floes, as Mr Godwin-Austen has done, an explanation of the isolated blocks of granite and lignite that have been recently found imbedded in the chalk of the south of England.

The marine fauna presents an exuberant display of sponge-growth (*spongia, ventriculites, siphonia, scyphia*, &c.), all less or more converted into flints; and leading to the ingenious speculation of Dr Bowerbank, that their function in nature is to induce the deposit of siliceous matter from the waters of the ocean just as the corals assist in the consolidation of its calcareous constituents. Foraminiferous organisms (*textularia, rotalina, dentalina, bulimra*, &c.) in countless myriads throng the waters, and drop their calcareous cases in such abundance, that more than the half of some chalk strata is composed of their exuviæ. It is now, and during the dawn of the tertiary period, that foraminiferous life attains its meridian—physiologically in hundreds of generic forms, and numerically in such abundance that

* Besides the cretaceous lignites of Europe, it is now known that the coal of Vancouver's Island and other American localities belongs to the same epoch. It is also more than likely that some coal-fields, now supposed to be oolitic, and several lignites now reputed of lower tertiary age, will yet be found to belong to the chalk formation.

thousands of miles of rock-matter (as the nummulitic limestone) owe their origin to the shell-like coverings of these

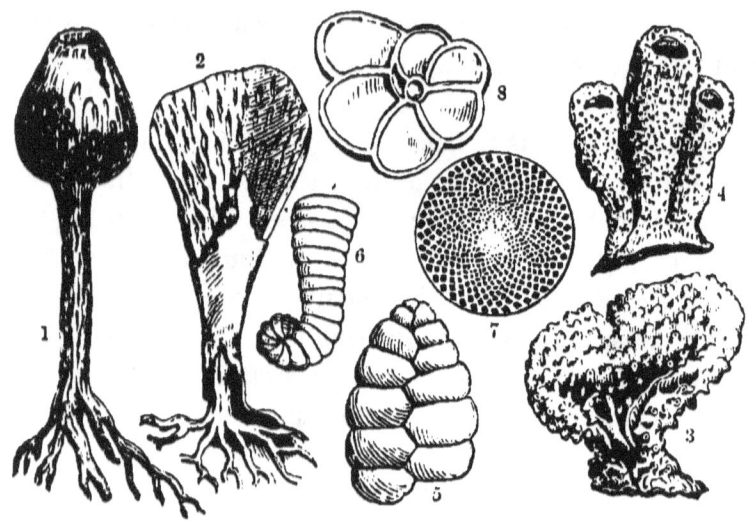

CRETACEOUS PROTOZOA.
1, Siphonia; 2, Ventriculites; 3 Manon; 4, Scyphia; 5, Textularia; 6, Lituola;
7, Orbitoides; 8, Rotalia.

the lowest of animated existences. Corals, though occurring in many genera (*parasmilia, trochocyathus*, &c.), are by no means so abundant as in the oolite; but star-fishes (*goniaster* and *oriaster*) and sea-urchins (*cidaris, diadema, salenia, galerites*, &c.) are obviously on the increase, and the beautiful preservation of the latter constitutes one of the most characteristic features of the chalk formation. Encrinite life is now drawing to a close, and in a few species of *pentacrinus, Bourgueticrinus*, and *marsupites*, carries down the descent to the solitary pentacrinite of the existing ocean. Annelids, like *serpula* and *vermicularia*, construct their tortuous tubes in profusion, and in such a marine medium barnacles (*pollicipes* and *scalpellum*) appear in greater force than during any former epoch. In crustacean life, the bivalved entomostraca are extensively represented by

Bairdia, cythere, cythereis, cytherella, &c., while the larger malacostracans abound in numerous generic forms, *myeria,*

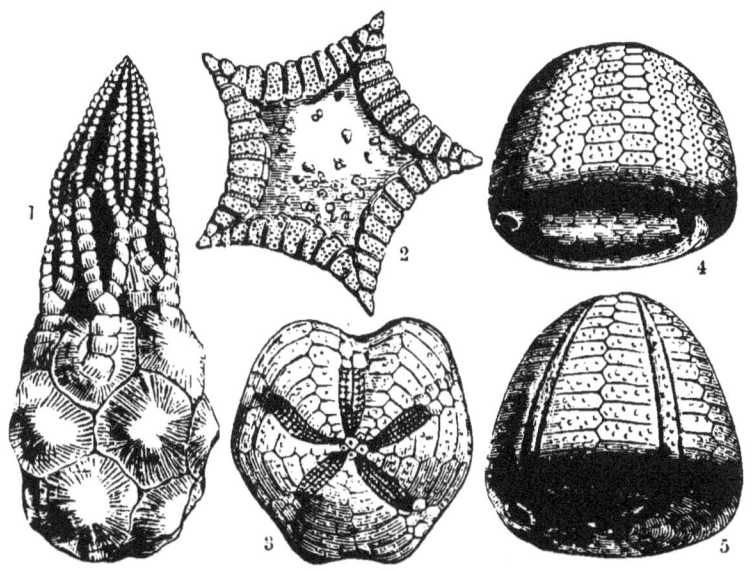

CRETACEOUS ECHINODERMS.
1, Marsupites; 2, Goniaster; 3, Hemipneustes; 4, Ananchytes; 5, Galerites.

pagurus, notopocorystes, &c., all more nearly approaching in aspect and function the crabs and lobsters of existing waters. And now the minute polyzoans or sea-mats weave their delicate tracery of network in a thousand forms (*flustra, eschara, diastopora, actinopora, idmonea,* &c.), spreading it over corals, dead shells, and crustaceans, as if their function had been to shroud in beauty the worthless and decaying wreck of the cretaceous sea-shore. The higher mollusca also appear in vast profusion—many of the oolitic genera having departed or being on the wane, while other forms peculiar to the chalk begin to make their appearance. The deep-sea brachiopods are represented by species of *terebratula, terebratella,* and *rhynconella;* the true bivalves by *inoceramus, lima, ostrea, pecten, astarte, cardium, tri-*

gonia, Venus, and many others whose specific forms are new and peculiar to the period; while gasteropods, like *natica,*

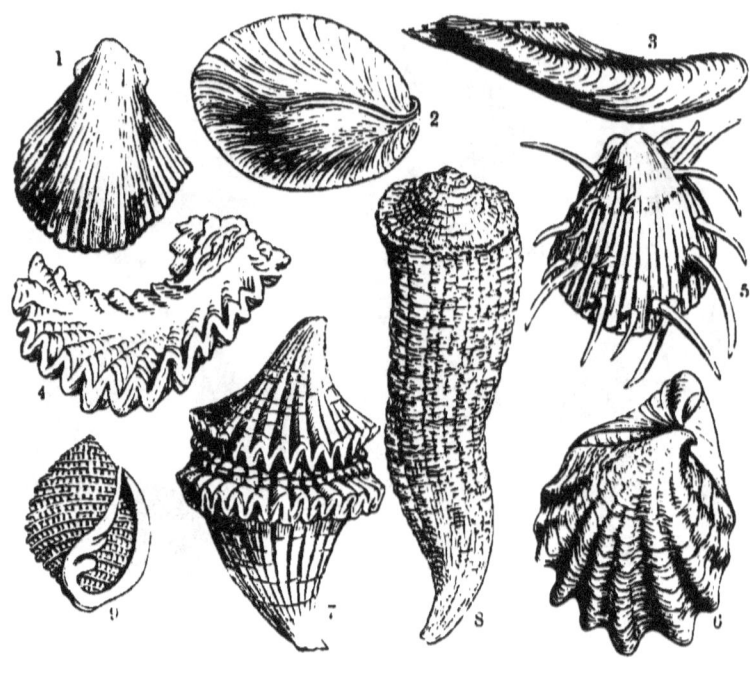

CRETACEOUS MOLLUSCA.

1, Pecten; 2, Terebratula; 3, Gervillia; 4, Ostrea; 5, Plagiostoma; 6, Inoceramus; 7, Radiolites; 8, Hippurites; 9, Cinulia.

littorina, cerithium, rostellaria, solarium, pleurotomaria, and others, mark a busy sea-shore of herbivorous and carnivorous activity. The cephalopods, though numerically fewer than in the lias and oolite, now appear in strange and fantastic forms. Hitherto the chambers of the shell-clad genera were either simply straight, like the orthoceras, or coiled on the same plane, like the nautilus and ammonite. Now, along with the nautilus and ammonite we have them bent like a hook (*hamites*), curved like the prow of a skiff (*scaphites*), incurved like a crosier (*ancyloceras*), curled like a ram's horn (*crioceras*), twisted round a straight axis, and tapering tower-like (*turrilites*), or in some other grotesque and simulative forms. This flush of generic

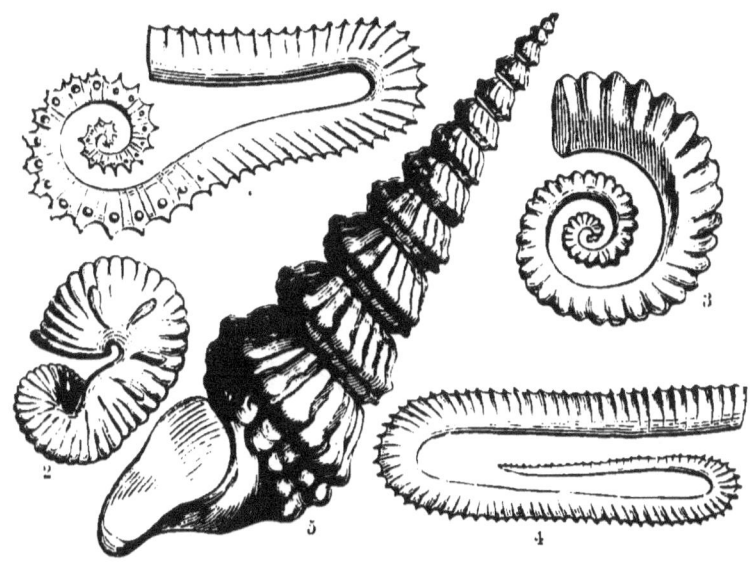

CRETACEOUS CEPHALOPODS.
1, Ancyloceras; 2, Scaphites; 3, Crioceras; 4, Hamites; 5, Turrilites.

type, and that on the eve of their decline, has given rise to many hypotheses; and by those who associate modification of form with the influence of physical conditions, obnoxious changes in the waters of deposit are supposed to have been the proximate causes of these curious and sportive shapes. It is true that an influx of fresh water into a marine area, or *vice versa*, is often attended by curious changes in the indwelling mollusca, and that new conditions of cultivation produce strange *sports* among the varieties of the gardener; but the forms of these cretaceous shells is too decisive and persistent to be otherwise explained than by the introduction of new genera, in obedience to some great but unknown law of creation. The fish life of the chalk period presents us with many of the old placoids and ganoids (the sharks, rays, and sauroids) of the oolite, but with new and peculiar genera of the same great divisions; while for the first time the ctenoids and cycloids, which are now the prevailing orders of ichthyic life, make their first appearance. Among the placoids, as indicating their fossil teeth, *ptychodus, hybodus, acrodus,* and *lamna* are the dominant genera; among the ganoids, *gyrodus, pycnodus,* and *macropoma;* while the cycloids show *osmeroides, hypsodus, saurocephalus,* and the like; and the ctenoids, the perch-like forms of *beryx* and *berycopsis.*

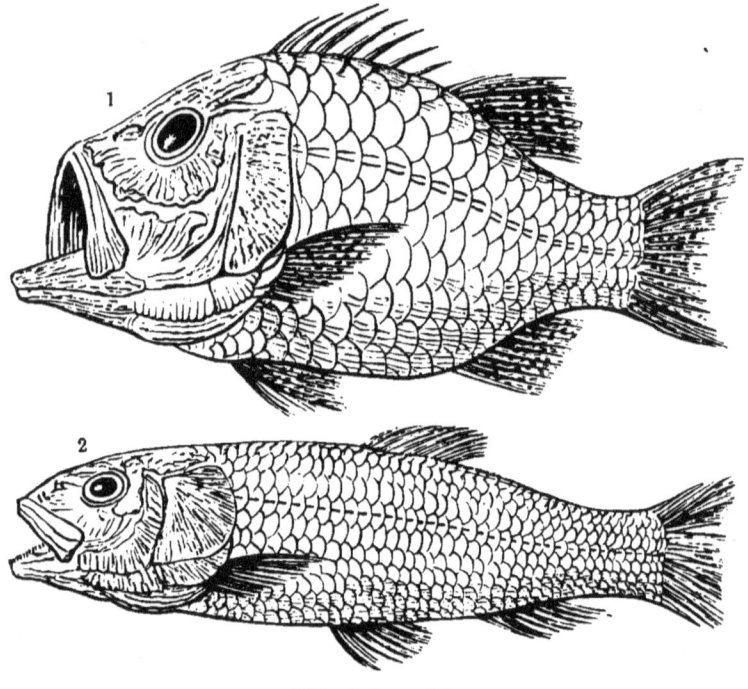

CRETACEOUS FISHES.
1, Beryx Lewesiensis; 2, Osmeroides Mantelli—MANTELL.

When we turn to the reptiles, a few species of *plesiosaurus* and *ichthyosaurus* still linger in the ocean; a solitary *iguanodon* represents the gigantic land-tribes; and *pterodactyles*, in lessening flocks, wing the sea-cliffs or skim the surface of the creeks and river-mouths. The crocodiles, lizards, and turtles are represented by several genera; but on the whole the meridian of reptilian life is past, and the huge and varied forms of the oolite are now extinct, or rapidly disappearing. Of birds and mammals the highly marine beds of the chalk have yielded little more than the merest indications (*cimoliornis*, bird-of-the-chalk-marl, &c.), but as these seem to point to the higher types of the rapacious birds and true mammals, we may rest assured of the existence of intervening orders, and look forward with hope to the discovery of their remains.

With the Chalk, which closes the long and prolific line of mesozoic life, we lose sight of many tribes, and families, and genera, but not, as is sometimes sweepingly asserted, of every species that up to that time had given character to the onward phases of vitality. The passage from the mesozoic to the cainozoic was as gradual as that from the palæozoic to the mesozoic, and if a break shall appear to exist in some districts, we cannot accept this as more than a mere local and limited phenomenon. The submergence of old lands, and the elevation of the sea-bed into new islands and continents, is a slow and gradual process; it is never cataclysmal save over the most partial and isolated tracts; and only in such tracts is there a chance of any genus or species being suddenly extinguished. As the gift of life is handed from generation to generation within certain limits of variety, so epoch passes it on to epoch within the wider limit of specific change, but this so imperceptibly that it is only after the lapse of ages the difference becomes apparent. Viewed at these wide intervals, the palæozoic flora seems essentially exogenous; endogens and gymnogens prevail in the mesozoic; and now the cainozoic is about to be characterised by the newer and higher manifestations of the exogens. In like manner

K

with the fauna : we rise (speaking in general terms) from a world of cold-blooded air-breathers in the palæozoic to cold-blooded air-breathers in the mesozoic, and from these again to the warm-blooded air-breathers of the cainozoic era. If fishes were the dominant vertebrates in palæozoic times, reptiles were undoubtedly so during the mesozoic; and now, in the cainozoic, the mammals (so feebly represented in the past) are about to assume the chief importance. The great march of life is not only ever forward, but ever upward. It is not merely that creation is concomitant with extinction, but the new creations are ever assuming more exalted ordinal forms of the same primal patterns.

THE RECENT.

CAINOZOIC SYSTEMS.—THE TERTIARY AND POST-TERTIARY.

HAVING passed the middle ages of the earth's history, whose life-species have all, or nearly all, disappeared, we enter upon an epoch whose forms insensibly graduate into those that are now our fellow-participators in the great progressional scheme of vitality. In other words, we approach the Cainozoic, or "recent-life period," which, though but as yesterday compared with the æons of the palæozoic and mesozoic, yet embraces a vast lapse of time, and is necessarily characterised by higher and still advancing forms. We say necessarily characterised, for though science can prove no causal connection between the physical and vital manifestations of the globe, the one set of changes so invariably accompany the other, that we are compelled to regard them as necessary concomitants. And yet, though concomitants in time, they may stand in no relation to each other as cause and effect, but be each an independent phase of that divine creational plan that is still evolving itself around us. We, who but dimly perceive the broken outline of the scheme, can only note the coincidence; those in after ages of higher intelligence may succeed in tracing the connection. But whatever that connection, it is now more marked and appreciable, and geologists can associate with almost every fluctuation of condition, a change in the accompanying aspects of cainozoic life.

It is now that the more complex forms of an exogenous flora are superadded to the endogens and gymnogens of the mesozoic, and in their more varied forms and higher utilities become not only a fitter ornament for a more varied surface, but a necessary sustenance for a higher and more diversified fauna. The herbs, and shrubs, and trees—the flowers, and fruits, and grains—all that can gladden the senses or satisfy the wants of man and his existing life-comrades, appear with the current epoch, and by their appearance again confirm that fitness that ever reigns between the organic and inorganic aspects of creation. In the animal world the advance is equally apparent, and in orders where no advance appears a thousand modifications present themselves. Among the protozoans the calcareous sponges for the most part disappear, their place being taken by those of a horny nature, while the foraminifera are culminating in size and complexity of configuration. The encrinites, with one or two solitary exceptions, have vanished from the waters; and the sea-urchins, so exquisitely preserved in the chalk, are reduced by several of their most beautiful and numerous families. Among the shell-fish the brachiopods dwindle to a few families, the true bivalves are still on the increase, the gasteropod univalves become dominant in genera and species, while the shell-clad cephalopods that thronged the mesozoic ocean in myriads, perish to a solitary genus. The crustaceans become less natatory and more ambulatory in their character, while the insects, so imperfectly preserved in the past, now throng every element—air, earth, and water—in apparently still increasing numbers. The placoids and ganoids, so long the only representatives of ichthyic life, are now on the wane, and the cycloids and ctenoids appear as the prevailing orders. Of the ichthyosaurs and plesiosaurs that whale-like ruled the ocean, of the megalosaurs and hylæosaurs that tenanted the plain and roamed the forest, and of the pterosaurs that winged the air, not a living trace remains. They are utterly extinguished, and their place is now filled by the crocodiles, lizards, turtles, and serpents of existing nature. The birds so scantily preserved (though largely indicated) in mesozoic strata, and the mammals represented only by a few insignificant marsupials, now assume the chief importance in the great vital scheme; and last, and highest of all, man himself enters on the stage of being as the crowning form of the current epoch.

To facilitate comparison, it is usual to subdivide the Cainozoic into eocene, miocene, pliocene, and pleistocene—that is, into its earliest, less recent, more recent, and most recent life-stages; but enough for our review to treat it in two great sections—the first, when land and sea had a somewhat different distribution from the present; and the second, when they had assumed, within the limits of an appreciable mutation, their existing arrangement. Adopting the familiar phraseology that designates the palæozoic as *primary*, and the mesozoic as *secondary*, we may regard the first section as *tertiary*, and the second as *post-tertiary*—ever bearing in mind that such distinctions are mere provisional aids to facilitate the comprehension of geological progression. It has been customary, no doubt, for certain geologists, generalising from limited tracts in Europe, to draw a bold line of demarcation between the chalk and tertiary—so bold that not a single species was regarded as passing from the one epoch to the other. This, like many of the early conclusions of the science, is altogether erroneous; and now, even in Europe, to say nothing of America, abundant passage-beds have been detected, showing in this instance, as in every other, that abrupt transitions are at the most merely local and limited phenomena. Assuming, then, that the life-forms of the Chalk pass insensibly into those of the Tertiary, even though in many European areas the cretaceous era was suddenly brought to a close by the violent displacement of the then land and sea, we yet discover a wide difference between the vital aspects of these respective epochs. In the northern hemisphere the tertiary seas still trend in an easterly and westerly direction—stretching diagonally through what is now Central Europe and Southern Asia, spreading over a large tract of Northern Africa, and covering in North America wide belts of the Southern States. Shut up from the northern currents that seem to have influenced the chalk seas, and exposed to those which, like the Gulf Stream, partake of a tropical temperature, the climate of the tertiary areas becomes more genial, and is, in the progress of creation, accompanied by a more exuberant flora and fauna. The seas, even in the latitudes of England, teem with southern forms; while the lands, clothed with a vegetation that finds its nearest analogues in the plants of sub-tropical regions, were tenanted by gigantic mammals which, like the elephant, rhinoceros, hippopotamus, tapir, lion, and tiger, now find their headquarters in the forests and plains of the torrid zone. Extensive lacustrine areas also appear in certain regions, as in Central France, and in their freshwater forms present, for the first time, a fauna but doubtfully and obscurely represented in former epochs.* In fine,

---

* With the exception of the estuarine beds of the Weald, and the doubtfully estuarine portions of the Carboniferous system, we are altogether ignorant of the fresh-water areas of the older epochs. Lake, river, and marsh must have existed then as now, each peopled by its own distinctive tenantry ; but of these forms we have not a single trace, and it is only as we approach the Tertiary epoch that a fresh-water fauna becomes known and appreciable. As we cannot believe in the total obliteration of ancient fresh-water deposits, so we hopefully look forward to important discoveries in this rich and varied section of vitality.

we have every type and feature of existing vitality; and the character of the period will perhaps be better indicated by a notice of the forms that have become extinct, than by any description of the whole, which still constitutes in a great measure the flora and fauna now flourishing around us.

Separating the early tertiaries—the eocene and miocene—from the pliocene and pleistocene, when, under the changing conditions of sea and land, the climate of the northern hemisphere began to assume a boreal character; we shall shortly glance at the more marked and peculiar aspects of this early period. Wherever we turn—whether to the clays and gravels of the London basin, or to the marls and gypsums of Paris, whether we restrict our review to the south of Europe, or carry it forward to the centre of Asia—we everywhere find in these earlier tertiaries abundant evidence of a warm-temperate, or even subtropical flora. Palm-like leaves and fruits, such as now flourish on the mud-islands of the Ganges (*flabellaria, nipadites, tricarpellites*), leguminous seeds of arboreal growth (*legumenosites*), twigs and leaves of mimosa, laurel, and other plants, whose congeners now find a habitat in southern latitudes, are thickly scattered through these strata. Nor are these the mere twigs and fragments of tropical forests, drifted from afar by gigantic rivers; for associated with the formation are beds of lignite or wood-coal, composed of kindred plants that must have flourished for centuries on the spots where their remains are now entombed.

And even if the flora gave no certain evidence of the geniality of the climate that then pervaded the parallels of London and Paris, the associated fauna would of itself establish the belief. Gigantic sharks and rays (*lamna, carcharodon, myliobatis*), such as now frequent the Southern Ocean, crocodiles and turtles (*crocodilus, chelone, emys*) in greater specific exuberance than is now known to the zoologist, tapir-like pachyderms (*palæotherium, anoplotherium*), akin to those of the Malayan peninsula and South America,

Restored Outlines—Xiphodon, Anoplotherium, Palæotherium.

and river-hogs (*hyopotamus, chœropotamus*), like those that now wallow in the mire of African rivers, have left their remains in thousands, testifying at once to the warmth of the climate and to the long continuance of conditions favourable alike to individual growth and to numerical abundance. An exuberance of pachydermatous quadrupeds, foreshadowing in their varied forms the solidungulates and ruminants of a subsequent era, is perhaps the most notable feature of the period; for, though the remains of whale, opossum, mole, bat, and even monkey, have been detected in the earlier tertiaries of Europe, the dominant impress of mammalian life over a larger section of the northern hemisphere was undoubtedly palæotheroid. In the forest, over the plain, and by lake and river-swamp, these curious creatures held supreme sway, simulating every form—sea-cow, tapir, hog, rhinoceros, ass, camel, antelope—and apparently performing every function now assigned to these later and diverse families. During the prevalence of the genial climate that then prevailed, we find not only an unusual flush of terrestrial life, but discover that the fresh-water lakes, the estuaries, and the seas, also teemed with many new and ascending forms. It is now that we detect in these marls the approximating species of our *lymneæ, paludinæ, plauorbes,* and other fresh-water shells; the terrestrial snails, *helix, pupa, clausilia,* &c., so slenderly represented in former epochs; and in the clays and limestones, increasing congeners of our marine gasteropods (spindle-shells, periwinkles, volutes, and cowries), *fusus, cerithium, natica, voluta, cypræa*—all assuming so recent an aspect, that the conchologist begins to rank them with living species, and to reckon the chronology of strata by the percentage of existing shells.* It is now, too, that the seas swarm with these foraminiferous organisms that attain their meridian in numbers, bulk, and variety, and give rise, by their myriad calcareous cases, to masses of nummulitic limestone

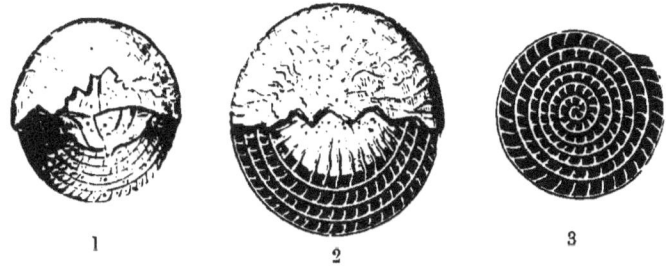

1, 2. Nummulites lævigata; 3, Section of do., showing its cells.

that rival in extent and thickness the limestones of former epochs, or the coral-reefs of the present day. The nummulitic limestone of the Old World, extending for thousands of miles, and many hundred feet in thickness, and the

* The terms eocene, miocene, &c., have reference to the percentage of existing shells contained in the different stages of tertiary strata, thus:—

    Pleistocene (most recent)   from 90 to 98 of living species.
    Pliocene   (more recent)   ,,   60 to 88   ,,
    Miocene   (less recent)   ,,   20 to 30   ,,
    Eocene   (dawn of recent) ,,   1 to 5   ,,

rivalling orbitoidal masses of the New World, are almost entirely the work of these many-celled foraminifera ; while thick and extensive strata of *tripoli*, and other siliceous earths, are wholly made up of the moulted shields of the minuter diatoms and infusoriæ. There is nothing more wonderful in nature than the magnitude of the stony masses elaborated by the lowest of animated creatures—the diatoms, foraminifers, corals, and other microscopic organisms. It seems as if an ordinance, that the nearer the vital approaches the physical, the less the organic is elevated above the inorganic, the more nearly they should resemble each other in the bulk and character of their lithological operations. And yet the elaboration of limestone from marine waters, by the merest vitalised speck of gelatinous matter, is a result that can never be mistaken for that of mechanical or chemical agency. The two things may approximate ; they can never be confounded.

Exuberant as the aspect of eocene life may appear, the march of creation is still ever forward. The physical agencies of nature are ever slowly but surely at work. Here the eocene sea is being gradually elevated into shoals and islands ; there the fresh-water lake is submerged, and its sediments overlaid by those of marine origin ; and here again volcanic energy gives birth to new mountain-chains, which interrupt the former currents of the air and ocean, and new external influences begin to prevail. The eocene gradually merges into the miocene, and the miocene into the pliocene. Old forms drop away, new ones begin to take their places, and a flora and fauna indicative of a more temperate climate begin to establish themselves over the latitudes that now constitute the middle regions of Europe, Asia, and North America. And just as we approach existing nature in time, so in space the flora and fauna begin to assume those distributive features that continue to characterise them more or less at the present day. The maples, planes, elms, willows, and other dicotyledonous trees contained in the middle tertiaries of Europe, bear the closest resemblance to those that still adorn her forests; and the elephants, hippopotami, rhinoceroses, bears, lions, and tigers of the Old World find their congeneric predecessors in the tertiary mastodons, mammoths, hippopotami, rhinoceroses, cave-bears, tiger-like machairodi, and camel-like merycotheres of the same hemisphere. In like manner the sloths, ant-eaters, armadilloes, and llamas of South America find their geographical prototypes in the megatheres, mylodons, glyptodons, and macrauchenes, so abundantly fossil in the upper tertiaries of that continent; while even in Australasia the kangaroo is preceded by the gigantic diprotodon, the lace-lizard by the megalanea, and the apteryx and emeu by the palapteryx and dinornis. As the miocene and pliocene epochs advance, the more and more do their fossil forms assimilate to those now peopling the same geographical regions, till the fossil may be said to graduate into the sub-fossil, and the sub-fossil into the species still existing.

In European tertiaries, for instance, we ascend from the eocene or palæotheroid age to the elephantoid or middle tertiary, and from this again to the later age of ruminants—antelopes, deer, and oxen. Connected as Europe has been with the rest of the Old World ever since the earliest tertiary epoch, we might naturally expect to find many species spreading indiscriminately over the other continents of Asia and Africa; but while this undoubtedly occurs, there are camel-like, giraffe-like, and antelope forms—*merycotherium*, *sivatherium*, *bramatherium*, and the like—peculiar to the tertiaries of Asia, which point to distinctive geographical distributions of life that obtained so early as the middle and upper tertiary epochs. And these curious forms—the huge camel-like merycothere and the elephant-antelope sivathere —suggest a peculiarity that runs through many of these tertiary mammals. Thus, while all the mammalian classes, with the exception of man, are less or more represented in the miocene and pliocene strata of the Old World, one feature that stamps the fauna of the period, and renders it noticeable even to unprofessional inquirers, is the vast amount of *intermediate or inosculating forms*. The horns of a ruminant with the proboscis of a pachyderm; the prehensile lip and dentition of a pachyderm with the light proportions of an antelope; the blending of horse, camel, and tapir; the inosculating of camel and giraffe—these and many other converging characters, appreciable only by the practised anatomist, are features that distinguish the tertiary mammals as a strange and peculiar fauna. Nor is it alone the more generalised physiological character, but their bulk is also in many instances a marked peculiarity of the period. The gigantic mammoths and mastodons, the huge hippopotami and rhinoceroses, the great cave-bears and cave-lions, the unwieldy megatheres and glyptodons compared with the existing sloths and armadilloes, the macrauchene compared with the llama, the trogontherium with the beaver, the diprotodon with the kangaroo, or the dinornis with the cassowary—all point to creational phases as the tertiary that have ceased to manifest themselves in the current era. The tusks of the mammoth have been found from twelve to fourteen feet (measuring along their outer curve), those of the existing elephant rarely exceed half that length; the fore-limb of the megathere would far outweigh the largest living sloth; the cuirass of the glyptodon would cover more than a score of armadilloes; the full-grown llama would make but a tiny calf to the macrauchene; and the emeu could walk beneath the stride of the extinct dinornis. This preponderance of bulky frameworks, and the number of intermediate forms that serve as connecting links between species now widely separate, are perhaps the most notable features of the tertiary fauna, and are highly suggestive to the physiologist, who, rising above mere description, strives to attain to the higher knowledge of creational method and law.

The diversified latitudes over which tertiary deposits are spread, and the difficulty of assigning a contemporaneity to strata containing few or no species in common, compels the palæontologist often to deal with the details of the respective areas rather than attempt a generalised expression for the whole. Enough for our outline, however, to remark that as sea and land approach their present configuration, the fossil flora and fauna begin in like manner to assume that distinctive impress which now characterises existing nature. As already stated, many of the Old World forms are unknown in the New; some of those that characterise the tertiaries of India are unknown in the strata of Europe; and only a few, and these during the earlier stages of the period, appear to have anything like a cosmopolitan extension. So also in the earliest or eocene stage the number of existing species are few compared with the extinct; this proportion increases in the middle stages; and as we rise to the uppermost deposits, it is often difficult to draw any specific distinction between the fossils they contain and the plants and animals that now flourish on their superficial areas. In the earliest stages the fauna of Europe was characterised by its palæotheres, anoplotheres, xiphodons, river-hogs, alligators, crocodiles, gavials, and turtles; in the middle stages these decline or die out, and deinotheres, mastodons, mammoths, camels, giraffes, cave-bears, lions, and hyænas take their places; while in the upper stages many of these decline, and mammoths, hippopotami, rhinoceroses, deer, wild oxen, horse, bears, and tigers become the dominant features. In like manner, when we turn to Asia, we can trace a similar ascent from the earlier stages, which contain many forms in common with those of Europe, to the middle stages, characterised by their numerous forms of elephant, sivatheres, bramatheres, camels, giraffe, lion, tiger,

Mammoth (Elephas primigenius). Mastodon.

monkey, crocodiles, and tortoises of enormous magnitude; and from these again to the upper stages, where the mammoth, rhinoceros, urus, horse, ass, and other creatures lead insensibly to the existing forms of that gigantic continent. In the same way it will be found with Africa, when geology has carried her researches further into that little known region; and so also it has been found in North

America, whose forms bear a wonderful parallelism to those of Europe; while in South America a similar gradation will yet be determined upward to those Pampean flats, whose pliocene clays and gravels have yielded those wonderful megatheres, mylodons, toxodons, glyptodons, mac-

Glyptodon, Megatherium (gigantic ground-sloth)—HAWKINS.

rauchenes, and other mammals, whose congeneric forms now people, in diminutive scale, the plains and forests and uplands of that exuberant continent. As with the larger continents, so with smaller and more detached areas. The marsupials of Australia have their forerunner in the gigan-

Dodo, Dinornis elephantopus, and D. ingens.

tic diprotodon; the wingless birds of New Zealand were

preceded by palapteryx and dinornis; and the still more gigantic æpyornis of Madagascar foreshadows the advent of the ostrich of Africa.

In the elimination of these successive fauna long ages must have passed away; and during these ages vast physical changes were necessarily effected on the terraqueous relations of the globe. In the northern hemisphere, some of the principal mountain-chains—the Alps, Apennines, Carpathians, and Himalayas—had been gradually assuming their ultimate configuration; and the large inland seas that had occupied the central latitudes of Europe, of Northern Africa, Middle India, and Eastern Siberia and China, had been elevated successively into shoals, lake, and island, swamp and dry land. Simultaneously with these terraqueous changes, the genial temperature that ushered in the eocene period of Europe and America began, stage by stage, to decline; the miocene was marked by more temperate manifestations; and ultimately the pliocene sank into a condition incompatible with the existence of the former flora and fauna. A cold, glacial, and barren period ensued, and under its rigours pliocene life in the northern hemisphere succumbed, and was succeeded by genera and species akin to those that now people the boreal regions.

This ungenial period, generally known in geology as the "Glacial," "Northern Drift," or "Boulder Clay" epoch, is lithologically characterised by its superficial mounds and masses of drift-sand and gravel, by thick tenacious clays, interspersed indiscriminately with water-worn blocks of all sizes, from mere pebbles to boulders many tons in weight, and by the polished, rounded, and striated surfaces of the subjacent rocks, as if they had been subjected to the long-continued friction of water or ice-borne material, and scratched and furrowed by the passage of the harder and heavier fragments. In Europe, Asia, and North America, down to the 44th or 42d parallel of latitude, and up to the altitude of 2000 feet, these appearances present themselves, and are inexplicable, unless on the ground of the gradual submergence of the northern hemisphere to that extent, and its subjection to a boreal climate which engendered glaciers on its hills, and drifted, during a brief summer, icebergs laden with rocky *debris* over its waters. The glaciers smoothing, rounding, and grooving the rocks of the higher grounds—the icebergs grinding their way through firth and strait, dropping their burden of mud, sand, and gravel on the sea-bed, or stranding themselves on its shores —complete the necessary arrangements for the production of the geological phenomena of the period. For ages the pliocene lands must have slowly subsided, each step gradually narrowing the boundaries of vegetable and animal life, and driving the surviving species, under the rigours of a deteriorating climate, to higher and higher regions. Race after race would succumb : first the more limited and local, next the more cosmopolitan, and ultimately few of the old flora or fauna would survive, except the more elastic in constitution, and those that had, step by step, retreated into more southern latitudes.

How long these conditions continued we have no means of determining in centuries, but, judging from the amount of denudation, the extent and nature of the heterogeneous deposits, as well as from the slow rate of elevation and submergence now going on in known regions, vast periods must have elapsed during the manifestation of this glacial epoch. At length the downward tendency of these northern latitudes comes to a close ; submergence stops and elevation begins. Slowly, and for long under a rigorous climate, the lands of Europe, Asia, and North America emerge from the waters. Glaciers still envelop the higher elevations; icebergs, summer after summer, drift over the waters; and the sea, attacking the soft emerging shores, re-assorts and re-deposits the sands, gravels, and clays of the older glacial epoch. By-and-by the deposits become fossiliferous, showing that the ocean was tenanted by shell-fish, seals, whales, and other creatures, whose habitats are now

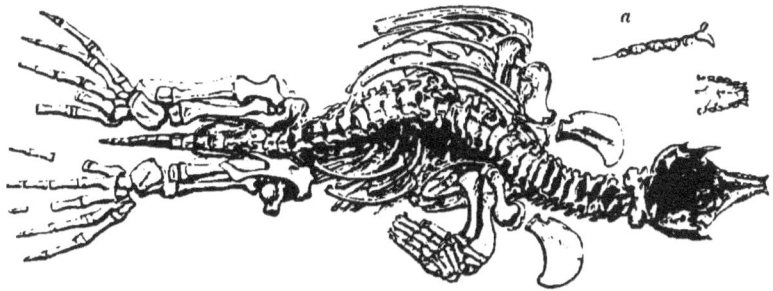

Skeleton of Seal (*Phoca vitulina*), from the Brick-clay of Suatheden, Fifeshire; *a*, dentition of do.

the icy regions of the arctic circle. Upward, still upward, the land emerges, evincing in its old water-lines and raised beaches the successive steps of its uprise, till ultimately the continents of the northern hemisphere assume, within appreciable limits of current mutation, the configuration and climatology they now present. As the continents

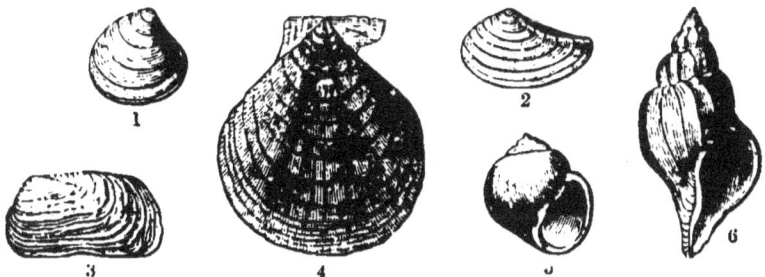

Boreal Shells in the Drift of the Clyde.—SMITH.
1, Astarte borealis; 2, Leda oblonga; 3, Saxicava rugosa; 4, Pecten islandicus; 5, Natica clausa; 6, Trophon clathratum.

emerge and the land surfaces augment, as new atmospheric and oceanic currents are established, and as the post-ter-

L

tiary epoch advances, the boreal races retreat farther to the north, some of the old pliocene families again return and spread over European latitudes, and other and newer forms, in the course of creation, begin to appear.

It is now the current era of geological history, whose vital record is the silts and marls of filled-up lakes, the alluvium of rivers and estuaries, the growth of peat-bogs and morasses, the stalagmite of fissures and caverns, and the tufa and ashes of volcanoes. In these superficial accumulations, which meet us at every turn, and are still in course of formation, every imbedded organism is fresh and familiar. With the exception of a few extinctions, the species yet flourish in the same latitudes, and the lists of the palæontologist become identical with those of the botanist and zoologist. The peat-bogs of Europe are replete with the mosses, grasses, willows, hazels, birches, firs, and oaks that still spread over our swamps, and adorn our forests. The tundras of Siberia, the jungle-soil of India, and the cypress-swamps of America, are in like manner composed of the plants now peculiar to these regions; and though in the course of geological change, local features may have varied, the main aspects of the Current Flora continue, zone for zone, and province for province, with little alteration or disturbance.

When we turn to the Fauna, the case is much the same. The most ancient lake-marls of Europe are thronged with *lymnea, paludina, cyclas, planorbis,* scarcely, if at all, distinguishable from those that now people our fresh-water ponds; and the marine shells of our estuarine silts and raised beaches—the mussels, cockles, oysters, periwinkles, whelks, silver-shells, and clams—with a few local variations, are identical with those that inhabit the surrounding seas. When we turn to the terrestrial fauna, the change, chiefly through the instrumentality of men, becomes a little more decided and apparent. The mammoth and mastodon, the Irish deer and urus, the cave-bear and hyæna, that seem to have roamed over Europe during the dawn of the post-

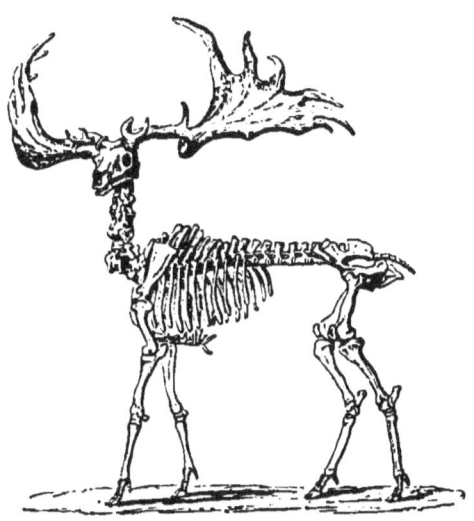

Megaceros Hibernicus, or Gigantic Irish Deer.

tertiary period, become extinct, though their congeners still flourish in Asia and Africa. As we ascend to later deposits, species, or, it may be, merely varieties of horse, ass, ox, deer, goat, sheep, bear, wild boar, wolf, and fox become the more frequent forms; and ultimately, in the more recent accumulations, the bones, whether of mammals, birds, reptiles, or fishes, become indistinguishable, even in variety, from those that are now our associates in the scheme of vitality.

And it is just in this palpable approach to existing nature that we begin to detect the earliest traces of the human species. First, and far back among the river-silts and peat-bogs and cave-earths, we discover his rude stone-implements and weapons, his tree-canoes, and the embers of the fires which he alone of all animals can either kindle or sustain. Side by side with these remains, occasionally lie bones of the mammoth, rhinoceros, and Irish deer; but whether these may not have been washed up, drifted, and re-assorted from earlier deposits, is a question not always easily determinable. However the question may be ultimately answered, one thing is certain, that just as the mammoths and mastodons drop away, and the horse, ox, goat, and sheep begin to spread over Europe in increasing numbers, so the traces of primeval man become more frequent and unmistakable. In all likelihood—nay, it is all but certain that over the plains and through the forests of the Old World man hunted the Irish deer and speared the mammoth, just as at a later period, and in the same region, he lassoed the wild horse and impounded the urus and buffalo. With regard to this subject, however—viz., the first appear- of man—much unnecessary discussion has taken place, and a great deal of uneasy tenderness been displayed. Like other events in geological history, we have at present no means of assigning to it a definite date in years and centuries. The time is merely relative, and all that science can safely do is to ascribe it to an early, though not to the very earliest, stages of the pleistocene epoch. Whether this was six thousand or sixteen thousand years ago, we cannot by any known data determine, though this much is evident, that the amount of change since effected on the physical and vital world, as well as the course of civilisation itself, would, at the current rate of progress, require for their elimination a much more extended period than is usually allowed.

And here it may be remarked, that while in these superficial accumulations we find frequent traces of primeval man—his stone-implements, tree-canoes, &c.—we rarely or ever discover the remains of man himself. Not a human bone has been detected, even in the valley of the Somme, where the flint-implements lie in thousands

—not a fragment where other fragments more slender and fragile occur in abundance. It is true, the search has yet been confined to a small portion of Europe; but the fact is somewhat significant, and forbids any attempt at generalisation till wider areas in Asia and America have been explored. Till this is done, and till bones and crania have been found and examined, it will be impossible to decide the ethnographic character of these early men, or to say whether they appeared in Asiatic, European, and American species, and consequently arose from various creative centres, or were merely time-distributed varieties of a single and one-created form. Geology, as far as the facts have been collated, gives no countenance to the idea of a plurality of creative centres. On the contrary, the sameness of the stone-implements, wherever they have been found, evince

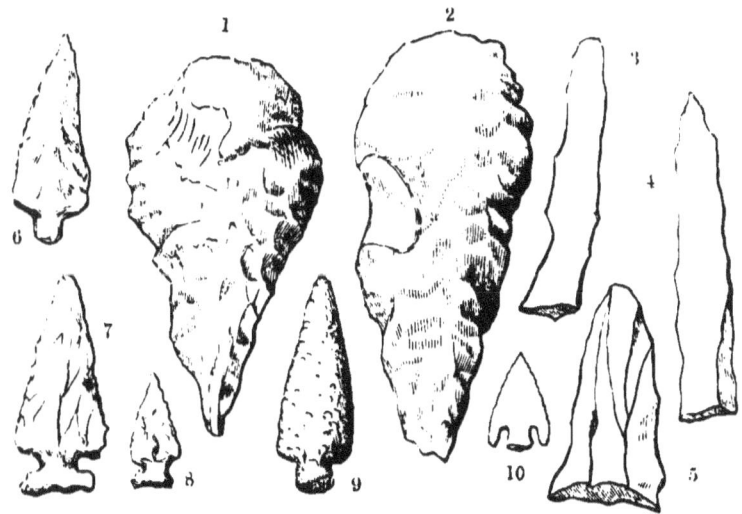

EARLY STONE-IMPLEMENTS.

1, 2, From valley of Somme; 3, 4, 5, England; 6, 7, 8, Canada; 9, 10, Scandinavia.

a similarity of idea—the same conception and the same design. Those, therefore, who, disregarding the unity of language, mental constitution, and religious sentiment of the human race, will still contend for several creative centres, must seek other corroboration of their hypothesis than is yet afforded by the discoveries and indications of geology.

As the pre-glacial passed gradually into the glacial, and the glacial into the post-glacial period; so the pre-human passes insensibly into the pre-historic, and the pre-historic into the historical ages. And even when the historical arrives, the record of our own race is often less certain in the hands of the historian than in those of the geologist. Geology by no means ceases where history begins. Vast physical changes have occurred since man first peopled the globe.* Some regions have been rising above the waters of the ocean, others have been sinking. Rivers have changed their courses; lakes and estuaries have been converted into alluvial tracts; and volcanoes have given birth to new mountain masses.

> "There rolls the deep where grew the tree;
> Oh, Earth, what changes hast thou seen!
> There, where the long street roars, has been
> The stillness of the central sea."

Of such mutations, history is altogether silent; and even where she speaks, her utterance is frequently of less value than her silence. The earth, however, pens and preserves with fidelity her own record: geology becomes her interpreter. As in the physical world, so also in the vital, important mutations have been effected, even within historical times. Many local removals of species and several general

---

* For an able and lucid exposition of the recent changes to which the earth has been subjected, the reader is referred to Sir Charles Lyell's *Principles of Geology*—a work which should be carefully studied by every one who would lay a logical and solid foundation for his geological knowledge.

extinctions have taken place, and this altogether apart from the effects produced by man's cultivation and domestication. The wild-boar, wild-ox, bear, wolf, and beaver have disappeared from Britain; and every century their tenure of Europe becomes more slender and uncertain. The dodo has become extinct in the Mauritius, the solitaire in Rodriguez, the æpiornis in Madagascar, the dinornis in New Zealand, the Phillip's Island parrot from Australia, and the rytina from the rivers and estuaries of Kamtschatka. And as with these, so it will shortly be with others whose circumscribed ranges are gradually being broken in upon by new conditions, imposed either by natural change or by man's progress and civilisation. The apteryx of New Zealand, the ornithorhynchus, echidna, and kangaroo of Australia, the mooruk of New Britain, the ostrich, elephant, and giraffe of Africa, the anrochs of Europe, the beaver and bison of America, the musk-ox of the arctic regions, and many others, look more like the residuary forms of the tertiary, than the advancing species of a newer era. And as with animals, so it has been and will be with many plants (the gigantic Wellingtonia, for instance, confined to a few narrow valleys in California); only we have been less observant of their mutations, and are merely beginning to note their specific restrictions.

As history has failed to note geological mutations and vital extinctions, so we ask her in vain for any evidence of new creations. No doubt, naturalists have now and then announced the "discovery" of a new species of plant or animal, but whether these were existing forms previously unnoticed, or new forms only recently introduced, the imperfection of history leaves us no means of determining. And yet, reasoning from our knowledge of the past, the appearance of new species must take place as infallibly as the disappearance of the old. So long as the energies of nature continue unimpaired, the balance of vital activity must be maintained. Even man's extirpations and modifications, extensive as they appear, are in a great measure counterbalanced by his introduction and wider distribution of the cultivated plants and domesticated animals in all their endless varieties. The scheme of Life is as progressive now as it ever was, and man himself is as subject to its laws as the meanest form he modifies. The pre-historic nomades of Asia, the stone-implement makers of Europe, and the mound-builders of America, have passed away, and are less known to us in their aspects, thoughts, and doings than their contemporary mammoths, great deer, and wild oxen. The temple-rearing, idol-worshipping races of Babylonia, Egypt, and Central America, have perished, and their characters are merely beginning to be revealed to us; while our more immediate predecessors, the Greeks, Romans, Celts, and early Saxons, have partaken of the same doom, and much of their history remains in doubt and obscurity. Thus, physical features, habits of life, modes of thought, social systems, and religious beliefs—all that renders humanity distinctive, and confers on it its highest attributes—have ever been as mutable and progressive as the phases of nature by which they are surrounded; nor do the realities of the present exhibit the slightest symptom of persistence and finality. As the palæozoic passed into the mesozoic, and the mesozoic into the recent; so the recent is pressing on to a future, that will be stamped by features—physical and vital, social and moral—peculiarly its own.

Supposing, then, that science could determine all the physical and vital conditions of the earth—in other words, could read her history up to the present moment—the question naturally arises, How far we are entitled, in the spirit of philosophy, to presume on what is yet to follow?

This brings us, in conclusion, to look at the earth's probable Future through her knowledge of her Past. As students of nature, we can no more refrain from this inquiry than we can cease to take an interest in her bygone history. The present is a mere evanishing point: yesterday it was the future, to-morrow it will be the past. Past, present, and future are but portions of one vast cycle of change; and could we determine with accuracy the rate of progress in the past, the future would be rationally computable. In the mean time our knowledge of world-history is far from perfect, hence our estimate of the future can assume at best little more than the character of speculation. Still, we are fairly entitled to hold that as the rocky crust has, under the operation of the physical forces, been the theatre of incessant change in the past, so it will continue to be subjected to similar mutations in the future. As we see no decline in the forces that operate, so reason refuses to admit a cessation of their results. Volcanic energy will shift its centres of activity; continents will be submerged; sea-beds be uplifted into dry land; climatic influences be altered; living races will succumb to obnoxious conditions; and new ones will appear co-adapted to these newer phases. As in the past the changes were always gradual and local, and the newer phases ever bore a certain appreciable relation to those that went before; so in the future we may rely on a similar gradation, and believe that the differences between the phases yet to be will never exceed those geology has discovered between two successive formations.

As with the physical, so with the vital forces. Age after age has been characterised by its own peculiar phases of vitality, and as we fail to detect any symptom of decline, so we may fairly presume that the future aspects of life will differ from that which now prevails, as that which exists differs from that which preceded. As the course has ever been to higher and higher forms, so the life of the future must transcend that of the present, as the present excels the past. Unless geology has altogether misinterpreted the history of this earth, and her teachings be no better than a fable and delusion, philosophy is chained to this conclusion. Could we discover the terms of the law that has regulated the evolutions of past vitality, we might approximate to some idea of its future forms ; but, ignorant of these terms, we can only rely on the upward progress of life, and believe that its newer phases will retain the same appreciable relations to the present that the present does to the age that immediately preceded. The great primal patterns—radiate, articulate, molluscan, and vertebrate—will ever remain the same : their modifications seem endless, their adaptations interminable.

THE LAW.

HAVING reviewed in detail the life-phases of the successive epochs of geology, we now proceed to a few generalisations respecting the advent and exit—the rise, progress, and decay—of specific vitality. In so doing, we shall endeavour to give expression to some of the leading laws which seem to have influenced Life since its first appearance on the terraqueous globe, believing that details are of themselves comparatively worthless unless we can co-relate and connect them into something like order and system. I am fully aware, where so much of our evidence is merely negative, and where more, perhaps, is still fragmentary and imperfect, that any attempt of this kind may be thought premature and perhaps presumptuous. But the law of our nature, like the law of creation, is ORDER ; and the mind instinctively groups and associates, and tries to connect effects with their causes, the moment it turns itself to any new field of research. And so, in Palæontology, these generalisations, however tentative and temporary, serve as centres round which to marshal new facts, and help to give consistency and interest to what might otherwise appear a mass of discordant and repulsive details. And granting that many of these generalisations may be set aside by future discoveries, so long as they are received in the same spirit in which they are submitted, they cannot retard the progress of research, by leading either to presumptuous dogmatism on the one hand, or to ungenerous illiberality on the other. They are submitted in the spirit of honest earnestness, more anxious to arrive at the expression of one plain truth than give currency to a thousand hypotheses, however brilliant and attractive. And yet, while the main value must ever be ascribed to inductive reasoning from facts, hypothetical promptings cannot always be ignored. They have their own value, and oftener than once has the road to truth been indicated by the fingerposts of hypothesis.

[Dawn of Life.]

As at present, so during all former life-epochs, the land and waters were tenanted by various families of plants and animals—these families exhibiting affinities and gradations even as plants and animals do now. It is true, that as we descend into the rocky crust we arrive at a stage (the metamorphic strata) when plants and animals do not seem to have existed; but on this point the evidence is merely negative, and Geology cannot say with certainty that Life was not coeval with the globe itself, though the presumption is, that organic being was not called into existence till about the dawn of the Silurian era. Nothing is gained by the assumption that it had a prior existence, and that every organism has been obliterated by the metamorphism to which the earlier strata have been subjected. We can only reason from what we know; and in the mean time the lowest Silurian or Cambrian rocks stand as the farthest verge to which Palæontology has pushed her discoveries.

It has been argued, no doubt, that as the vertebrate animals seem to show an ascent through the geological periods from fish to reptile, from reptile to bird, and from bird to mammal, so the invertebrate may also obey a similar law of development, from the simpler protozoa up through the radiata, articulata, and mollusca. As every class, however, of the invertebrata is represented in Silurian strata, we must, according to this hypothesis, seek for the commencement of the simpler forms, stage by stage, further back among the Cambrian and metamorphic rocks. Such a belief would carry the commencement of life immensely further back in time—as far back, indeed, before the Silurian period for the development of the invertebrates, as from the Silurian to the tertiary for the vertebrates; but as the same ratio cannot possibly be applied to two sub-kingdoms so entirely dissimilar, the idea of a long pre-Silurian cycle of invertebrate life, however plausible, has really little in fact to recommend it to our acceptance. We by no means argue for the restriction of life to the Cambrian period, but we must have something more certain than fanciful analogies to carry our convictions any distance beyond these strata. And even the evidence of Fossil life itself is greatly in favour of the belief that at this stage we have reached, or all but reached, the dawn of organised existence. As we ascend in the geological scale, we find life increasing and spreading, stage by stage, into newer and higher forms; and as we descend we find it decreasing and narrowing to simpler and lowlier aspects; and surely we are justified in the inference, that in the few scattered organisms of Cambria, we have all but attained the ultimate limit of vitality. Were matter and life co-dependent we might reasonably argue for their co-existence; but as matter can exist without the manifestation of vitality, and as life appears only in subordination to the material forces, so the one may have existed for ages without necessarily implying the presence of the other.

And, further, if untold epochs have been spent in the evolution of life from its earliest to its present aspects, it is equally conceivable that cycle after cycle may have rolled by in the elimination of the purely material structure of the world before it seemed to the Divine mind a fitting habitat for the plants and animals with which He had predestined to adorn its surface. Life is a measured and restricted gift; it is adjusted to, as well as governed by, external conditions, and it is only in harmony with what we know of Nature's progress to believe in a long Azoic period, during which these external conditions were undergoing the necessary preparation and arrangement for the advent of Vitality. At present all our ideas of Life are associated with a globe superficially composed of land and water, surrounded by a breatheable atmosphere, lighted and warmed by a genial sun, and subjected to ever-acting physical and chemical forces; and while we firmly believe that independent of these great cosmical conditions Life could not exist, we may surely be permitted to presume that it has been their unfailing accompaniment from the earliest moment they were harmoniously established, and will continue to be so while that harmony remains undissolved.

[Origin of Life.]

Starting from this point, we may fairly inquire, How and by what means this earth became the "procreant cradle" of organised existences? Was it by some process of secondary causation, or directly and at once by the fiat of the Creator? Alas for the impotence of science, and the scope of our finite intelligence! Science cannot even indicate the line of inquiry—our highest philosophy is the humble recognition of the fact. The chemist and the physiologist may resolve the vital organism into cells, and granules, and nuclei, but here their efforts stop: they cannot endow these cells and germs with life, or cause them to assume the lowliest form of vegetable or animal existence. The "slime that mantles o'er the stagnant pool"—the simplest arrangement of cell-growth that spreads over the surface of the decaying rock, are results beyond the proudest achievements of science. And even could we in any way connect these manifestations of life—lowly as they are—with the subtle agencies of heat, light, and electricity, this would be only bringing us a little nearer, but not within the precincts of that mysterious shrine which science may not unveil, and before which the proudest philosophy can only humble itself and adore.

It may be a law of nature that inorganic matter, in certain conditions and under the influence of certain forces, shall assume an organic form, but of the operations of such a law, or of the forces which obey its behest, human knowledge stands in utter ignorance. It may be, as suggested by Professor Owen, that "if it be ever permitted to man to penetrate the mystery which enshrouds the origin of organic force, it will be, most probably, by experiment and observation on the atoms that manifest the simplest conditions of life," but in the mean time the lowly monad stands as unrevealed as to its origin as the lordly man, or as the still more subtle elimination of mental phenomena. We know something of the nature and functions of vitality—its order, and operations, and increase—but of its origin we know nothing. In vain have physicists experimented to associate vital manifestation with electrical action : most unsatisfactory the evidence for what naturalists, in ignorance of the phenomena, have been pleased to designate spontaneous generation ! This present ignorance, however, can form no plea for the absence of future effort ; everything unknown is not to be held as a miracle. On the contrary, where natural science, under the direction of proper methods, has already done so much towards the elucidation of Life in all its aspects and operations, philosophy may surely be permitted, humbly and reverently, to inquire into its cause and origin. The distance that separates the Uncreated from the created is no doubt inconceivably great, but who shall fix the " Hither shalt thou come but no farther" of the legitimate efforts of the human intellect?

[Uniformity of Type and Pattern.]

At whatever stage the first creation of plants and animals took place, the same types and patterns have ever run throughout the whole—analogous functions have had to be performed; and the various biological provinces of every epoch have been peopled, even as they are now, partly by identical and partly by representative species. The earliest sea-weed that floated in Silurian waters spread its broad leathery lobes, and grew and multiplied precisely like the sea-weeds of existing seas; the shell-fish were constructed after the same types, and so similar in many respects, that only the practised conchologist can detect their specific distinctions; the fishes of the old red sandstone, strange as they may appear, find their analogues, bone for bone, and organ for organ, in existing fishes; while the reptiles of the Oolite and Weald present, in their gigantic skeletons, nothing that the comparative anatomist cannot readily identify with the homologies of the vertebrate pattern. And not only the structural *plan*, but the structural *composition* of each great type has been equally persistent and normal. The mucilaginous sea-weed for ever leaves its faint impress devoid of cortical integument, the terrestrial stem has ever constructed its woody pillar of carbon, the equisetum elaborated its varnish of silica, and the pine-trees distilled their resins and amber. The corals of Siluria, like those of the existing Pacific, consist of carbonate of lime; the fossil and the recent tooth have alike their dentine, osteo-dentine, and enamel; and had the parts been less evanescent, or were chemistry more subtle to detect, bone, cartilage, and flesh — hair, wool, and horn — scale, feather, and claw, would be found to have been ever built up of the same elements, and, family for family, in the same proportions.

So far as fossil evidence goes (and by that alone can we be guided), the plants and animals of the ancient world, though differing widely in genera and species, were neither "abnormal" nor "monstrous," but both in point of size, and form, and structural adaptations, were very much alike to those of the present day. So much so, indeed, that could we recall them to mingle in the busy scene of life around us, they would neither startle us by their appearance, nor alarm us by their habits, one whit more than the existing Flora and Fauna of distant and different regions. The great types remain the same throughout all time and space; and though the modifications have been innumerable, these modifications, even in their aggregate, have never amounted to an obliteration of any important primal distinction. Acrogenous, endogenous, and exogenous—radiate, articulate, molluscan, and vertebrate, range side by side as distinctly now, each within its own typical idea, as when they first clothed the land and peopled the waters. The relations of a mathematical line, or the unions of a chemical element, are not more fixed and certain than the relationships of structural organisation. This organ or that organ may be modified for the performance of some special function, and this modification may imply a certain amount of variation in all the co-related parts; still the prime typical conception remains distinct and essential under every condition of space, and through every progressional mutation of time.

"Nor is it only the *plan* of the great types," says Pro-

M

fessor Agassiz, " which must have been adopted from the beginning, but also the *manner* in which these plans were to be executed : the systems of form under which these structures were to be clothed, and even the ultimate details of structure which, in different genera, bear definite relations to those of other genera ; the mode of differentiation of species, and the nature of their relations to the surrounding media, must likewise have been determined ; for the character of the classes is as well defined as that of the four great branches of the animal kingdom, or that of the families, the genera, and the species. Again, the first representatives of each class stand in definite relations to their successors in later periods; and as their order of appearance corresponds to the various degrees of complication of their structure, and forms a natural series closely linked together, this natural gradation must have been contemplated from the beginning." In other words, the evolution of life in all its successive phases is but the realisation of a pre-determined plan ; and whether primary or secondary means be employed in its enactment, the harmonious unity of their action implies alike omnipresence in space and omniscience in time.

[Function.]

As to function ; earth, air, and water ever seem to have had their varied tenantry. Burrowers, creepers, runners, leapers, fliers, floaters, and swimmers, make their appearance in every epoch. Simple and lowly they may be, yet still in their respective grades perfect, and fitted by the nicest organic adjustments at once for the functions they had to discharge and the element they were destined to inhabit. From these organs we also clearly perceive, that some families were designed to feed on vegetables, others to prey on flesh; that some were formed to roam at large for their food, others to find it by parasitic attachment; while many, like the crustacea of the lower old red, the sauroid fishes of the coal period, and the reptiles of the lias, became the scavengers of their respective times, and lived on the decaying garbage of the river-bank and the muddy seashore. The functional performance of each great class, as well as of the life of each great geological epoch, has ever been, within its own limits, a complete and independent system. A world of shell-fish—littoral and deep-sea, sedentary and vagrant, phytophagous and carnivorous—existed in the earliest waters. The gigantic sauroid fishes of the palæozoic were the functional representatives of the secondary reptiles; the secondary reptiles, in their marine ichthyosaurs and plesiosaurs, their estuarine teleosaurs and steneosaurs, their terrestrial hylæosaurs and megalosaurs, and their aërial pterosaurs, were respectively the whales and dolphins, the crocodiles and gavials, the elephants and tigers, the bats and the birds, of their period. At every stage of time, and under every type of life, analogous functions have been unerringly discharged. Herbivorous, insectivorous, carnivorous, and omnivorous, are attributes alike of the fish, the reptile, the bird, and the mammal; walkers, swimmers, and fliers, with powers more or less restricted, have ever occurred within the same great classes.

In the interdependencies of existence demand has ever pressed on supply, decay trodden closely in the wake of reproduction, and suffering been commensurate with enjoyment. An ideal Cosmos of painless beatitude is a dream and delusion. Pain and death are stamped on the earliest records of life. From the beginning the flesh-eater has preyed on the plant-eater, and the weak have ever succumbed to the strong, even as they do now. The struggle for existence commenced with its gift; and the reign of death was inaugurated by the enjoyment of life. Constructed as Nature is, this seems part and parcel of her plan, and the means by which the equipoise and balance of vitality is maintained. The larger and more abundant plant-feeders, ever pressing on the means of subsistence, are held in check by the comparatively smaller and scantier flesh-eaters; and, so far as man can comprehend, it is only by some such compensatory system that the greatest happiness of the greatest numbers can be maintained. Besides, subjected as life is to the inevitable laws of a material world, it must, for its own comfort, learn to accommodate itself to the circumstances by which it is surrounded; and so, under this view, the accident and reminiscence of pain become an institution for the animal's own benefit and protection. What pleases will be pursued, what pains will be avoided; while the excess of force which destroys is, for the most part, mercifully accompanied by insensibility and unconsciousness. Life, like the world it inhabits, is after all but a system of re-agency and compensation; and in all our reasonings on the question of Pain and Death we should ever remember that "He who tempers the wind to the shorn lamb" may have so ordained, that to the various grades of organisation suffering and the terror of death should be merely comparative, and that their intensity should be felt only where pain becomes the penalty of the infringement of the eternal law of right and wrong.

[Distribution.]

With regard to distribution, we also perceive that from the beginning different areas have been peopled partly by identical and partly by representative species. In the Palæozoic epochs, the marine areas being apparently wider and more conterminous, the same species had, perhaps, a more extensive range; but even then there was no universal uniformity of life—a thing as incompatible with the habits of the creatures themselves, as uniformity of climate is with the form and motions of the globe. Plant the first germ of life on whatever spot you may, this act of creation has always some relation of fitness to external conditions; and as universal uniformity of condition is a thing unknown either in time present or in time past, so we may rest assured that universal uniformity of Life was a feature that never entered into the scheme of creation. Tropical and temperate, low-lying and elevated, littoral and deep-sea, have ever prevailed as distinctive areas, impressed by different physical conditions, and requiring for their tenancy orders and families equally distinct in their habits and organisation. No doubt certain animals, in consequence of their periodic migrations, have a much wider range than others, but even this is fixed and ascertainable, and these migratory races, at the present day at least, are comparatively few. Applying this rule to the past, we may believe in the migrations of certain extinct races,* but these always within definite limits, and, race for race, each over its own appointed area.

From the first to the last, variety and complexity are unmistakably stamped on all created forms; and this variety manifests itself not only in the structure of the creatures themselves, but in their general *facies* at successive epochs, as well as over the different areas they were meant to in-

---

* The migrations of living animals are comparatively well known; the migration of extinct races is altogether wrapt in obscurity. And yet we may believe that the mastodons and mammoths had a wide range from south to north, that some of the tertiary birds dipped their wings alike in temperate and arctic waters, and that many of the secondary fishes were, like the salmon and sturgeon, anadromous—now frequenting the sea, and now returning to the river under the periodic impulse of reproduction.

habit. Infinite variety of structure, infinite variety through time, and infinite variety over space, seem leading characteristics of the Creator's scheme; and we err in our interpretation when we try to establish for the primeval world a law that has ceased to operate in the new. It is true that as we ascend in time, from higher to higher forms, the areas of specific distribution seem to become more and more circumscribed, and such a limitation would only accord with the idea of increased diversity of species as dependent on more localised varieties of food, climate, and other external conditions; but even on this point we must exercise great care and discrimination. In the present continents we trace, in some measure, the outline of former seas and lands; but more than two-thirds of the earth's crust are covered by water, and hide from our research the continuations of systems, a knowledge of whose extent is indispensable to the solution of the problem. It is better, then, to shape our inferences in accordance with what we know of existing nature, and believe in variety and distribution of species from the first, in various centres of creative manifestation, and in a process of local extinctions and creations which necessarily prevented universal uniformity of Life during any of the geological epochs.

[External Conditions never Uniform.]

It has been argued, no doubt, that in the primeval epochs our earth, in virtue of its internal heat, enjoyed a higher and more uniform climate, and was consequently peopled by a more uniform flora and fauna. This argument, like many others of the earlier geologists, seems altogether without foundation. Granting the existence of a higher internal temperature in pre-vital times, and admitting its influence on the surface, we are still without a shadow of evidence that this interior heat has exercised the least perceptible effect on the climatology of the globe since the deposition of the fossiliferous strata.* On the contrary, all that we know of the nature and thickness of pre-Cambrian sediments, and all that we have learned of the Cambrian rocks themselves, preclude the supposition of such an influence beyond the most infinitesimal degree, and compel us to believe that the physical conditions of life have been much the same throughout every period of its existence. It is true that during the Carboniferous period, the Oolite, and the earlier Tertiaries, certain latitudes of the northern hemisphere appear to have enjoyed a more genial climate than they do now; but the explanation of this we are to seek in the varying distributions of sea and land, the existence of warmer oceanic currents and other geographical conditions, rather than in any perceptible influence derived from the earth's interior. Nay, as we have warmer and colder regions *in space*, in virtue of the earth's relations to the solar system, so we are inclined to believe we have had warmer and colder periods *in time*, in virtue of some great but unknown cosmical law. The existence of the Glacial or boulder epoch over the greater portion of the northern hemisphere (say up to the 40th or 42d parallel of latitude) is now admitted on all hands; and as we cannot entertain the idea of

* According to Fourier, Hopkins, and other physicists, the internal heat of the globe, which increases at the rate of 1 degree for every 60 feet of depth, does not at present affect the mean superficial temperature more than 1-20th of a degree; and to have had any sensible effect on external climates— say to the exent of 10 degrees—this interior heat must have been two hundred times its present amount. At that rate the melting point of lavas would have been reached at a depth of 580 feet, instead of 116,000 feet as presently estimated, and all the deeper seated strata must have been fused or rendered crystalline—a condition in which they do not occur even to the depth of 30,000 feet, as many of the Cambrian and Silurian slates are merely hardened and cleaved, but in no degree metamorphic.

cataclysmal irregularity in Creation, so we are led to infer the prior occurrence of such glacial periods at determinate times and over determinate areas. The existence of such glacial recurrences has been surmised by several geologists as characterising the periods of the old red sandstone and Permian;* and we may venture to extend them to other systems as probable features of a great cosmical plan.

Thus, looking at the Cambrian strata of the northern hemisphere—their angular grits and conglomerates, their extreme paucity of fossil forms, and other features—we are at once reminded of the action of ice and the presence of ungenial conditions. This is followed over the same areas by the more genial and exuberant period of Siluria; which is in turn succeeded by the old red sandstone, whose grits and bouldery conglomerates, as well as paucity of vegetable forms, once more suggest the recurrence of colder influences. Following the old red we have the exuberant flora and fauna of the coal period, again to be succeeded by the scanty life-forms and grits and conglomerates of Permia. Again, the trias and oolite of the northern hemisphere are characterised by life-forms that betoken warm and genial conditions; while the chalk that succeeds imbeds water-worn blocks of granite and lignite, which would seem to imply the presence of ice-drift and deposit in seas that were open to boreal influences. Next the early tertiaries occur over the same areas, marked by plants and animals that indicate a warm and genial climate; and this in turn gives place to the well-known glacial or boulder-drift epoch; once more to be succeeded by the milder influences of the post-tertiary or current era. Throwing these recurrences into diagrammatic form, we appear to have had an alter-

---

* Mr Cumming, in *History of the Isle of Man* (1848); Mr Goodwin-Austen, *Geological Journal* (1850); Professor Ramsay, *Ibid.* (1855); and the author, in his *Advanced Text-Book.*

nation of colder and warmer cycles over the northern hemisphere, at least; and if such has really been the case, we must seek the explanation not in revolutions and cataclysms, but in some fixed and continuously operating law.* Whether the phenomena may depend on causes operating on and within the globe itself, so as to change the axis of rotation, or whether it may not more

* This idea of colder and warmer cycles as affecting the northern hemisphere was indicated some years ago to the Literary and Philosophical Society of St Andrews. Since then the author has endeavoured to establish the fact, partly by the character and composition of the rocks of the colder periods, and partly by the nature of their fossil contents. Much, however, still remains to be done, and he would earnestly solicit the attention of geologists to the subject, and this altogether apart from the cosmical causes to which the recurrences may be due. On this latter aspect of the question some discussion took place in the *Athenæum* of 1860, on the suggestion of Colonel James of the Ordnance Survey, that former changes of climate may be due to changes in the inclination of the earth's axis, brought about by alterations in the crust that gradually affect the centre of gravity. Whatever the cause—whether it is to be sought for on or within the globe itself, or in purely astronomical influences—this is not the place to discuss; but most unmistakably the gradual uprise of land that is now taking place in the arctic regions, the shifting of volcanic arcas in the northern hemisphere since the tertiary period, and the approach and departure of the boulder epoch over the same latitudes, all point to the operation of some determinate law of secular succession. May it not be, that in the periodicity of this law we may yet discover the key to the expression of geological chronology in years and centuries?

likely depend on forces purely astronomical, are questions that lie beyond the scope of the present Sketch. Enough for our purpose to have indicated the probability of such recurrences, and derive therefrom the conclusion that the conditions of life have been very much the same through all geological periods—successively varying in different areas, but never presenting, any more than they do now, a universal similitude, and that least of all through the influence of the earth's internal temperature.

[Introduction of New Life-forms.]

As each geological epoch is characterised by its own peculiar plants and animals, the question naturally arises, Whether these are independent creations, or whether there is in nature some law of development by which, during the lapse of ages and under the change of physical conditions, the lower may not be developed into the higher species, and the simpler into the more complex? On this topic much has been said and written, but after all, geology is not in a position to solve the problem of vital gradation and progress. It cannot tell, for instance, why trilobites should have flourished so profusely during the silurian epoch and died out before the deposition of the oolite; why chambered cephalopods, like the ammonite, should have come to their meridian, as it were, during the liassic era; reptilian life during the oolite and chalk; or why mammalian development should have been reserved to the tertiary and current epochs. It cannot explain why the palæotherium should not continue to inhabit the same forest with the tapir of South America, or the ichthyosaurus gambol in the same waters with the alligator of the Amazon. It can discover no physical condition in the oolitic seas to have prevented the continuance of trilobites; nothing in the geography or climate of the coal period to have prevented the huge terrestrial reptiles of the Weald from browsing on its vegetation, or marine species, like those of the lias, from preying on its fishes. The appearance and preponderance of certain races during certain geological epochs is a problem which lies as yet beyond the solution of science. That this succession occurs regularly as regards time, space, and biological sequence, we clearly perceive; but how, or by what means of causation, we are altogether unable to determine. We can often trace the extinction of races to a change of external condition; and as vitality is endowed with a certain amount of elasticity and adaptability, we may account for modifications within the limits of what naturalists term *varieties;* but we appeal in vain to physical conditions for the first introduction or creation of species.

In the PAST LIFE of the globe we only see dimly and broadly the outline of a great scheme of gradation and progress—a progress on which we may rest as a matter of FAITH, but the terms of whose LAW lie far, as yet, beyond the grasp of exact scientific demonstration. In vain we turn to "external conditions" and "unlimited time;" to the doctrines of "embryology" and "morphology;" or to "natural selection in the struggle for existence." These are oracles to which theorists have often appealed, but they fail, as yet, to utter an intelligible response. That each of them has some portion of the mystery in keeping, all the tendencies of modern science do, no doubt, appear to indicate, but how much, and in what order of connection, our highest determinations are little better than a train of ingenious guess-work. As far as geological evidence goes, all the great types of life began simultaneously and independently. All the subsequent introductions of new genera and species are but modifications of these types; but how, or by what process they were modified, science cannot tell, any more than it can account for the creation of the type itself. This much we know—if the geological record is to be trusted—that age after age new forms of life have made their appearance, differing in what naturalists would term *generic* and *specific* aspects, but still bearing to the great primal patterns, and to each other, certain definite and appreciable affinities; and as we are not entitled to place vital phenomena any more than physical phenomena beyond the pale of natural law, we are bound, in the spirit of philosophy, to seek inductively for the causes of these successive introductions. In the whole world around us we see nothing but the activities of secondary causes; and though Reason has yet failed to detect the mode in which new life-forms are produced, Faith may surely be allowed to believe in their genetic connection by some continuously operating law. To such a law science can give no satisfactory expression; and in the mean time the idea of New Creations is, if not the most philosophical, at least the most prevalent belief, just as it is the most convenient term, perhaps, whereby we can describe the phenomena in question. Instead of the term "new creations," some palæontologists, with a view to avoid an opinion, make use of the phrases, "the first appearance," and the "introduction" of new races. Little, however, is gained by this evasion. If new species do enter upon the stage of being, and we cannot explain how or by what process they come, then they are to us, to all intents and purposes, *new creations.* It may not be a new creation in the sense of a direct and miraculous interference on the part of Creative Power; but it is the equivalent of creation through the operation of a law defined by the prescience of a Creator, and producing its results at determinate times, over determinate areas, and always with a determinate relation to pre-existing vitality.

[Extinction and Creation of Species.]

In adopting the terms "extinction" and "creation," we must not fall into the common, but mistaken, notion, that the Flora and Fauna of one period were utterly extinguished before the commencement of the next. There are no such extinctions and re-creations in nature. Just as the physical change from one formation to another was sudden or gradual, so a less or greater number of genera and species passed from the older to the newer epoch. In some localities the change was sudden and entire, in others it was gradual—so gradual that we can hardly trace the line of demarcation. Take, for example, the old red sandstone of the British Islands. What a vast difference between the fauna of Siluria proper, and that of the old red sandstone of Caithness! The break seems decided and impassable, and yet when we turn to Forfarshire and Lanark we find silurian genera and species passing up into the old red sandstone and completing the continuity, which, to a Caithness geologist, would have seemed to be entirely rent asunder. Again, what a marked difference between the fauna of the Forfarshire and Caithness beds — between that of the Hereford sandstones, and that of the limestones of Devon! and yet when we pass to the old red sandstone region of Russia we find these different stages fused and equalised into one homogeneous Life-system. What was broken up into different stages by physical irregularities in the area of Great Britain was left to evolve itself gradually and continuously in the region of Russia. We must examine more and know more before we hasten to such sweeping conclusions as general extinctions and creations; and the more we examine and know, the more we become convinced that geology cannot point its finger to a single break in the great evolution of vitality, any more than it can point to a moment's cessation in the physical operations of nature. On the contrary, geologists now know that a considerable number of species always pass from one formation to another; and such terms as "passage-beds," "Cambro-Silurian," "Siluro-Devonian," &c., sufficiently express their conviction that the outgoings and incomings of life-forms are inseparably interwoven into one gradual and continuous sequence.

The whole of our groups and formations are merely successive stages in one great system or Cosmos—the minor stages imperceptibly graduating into each other, and the amount of progress becoming apparent only after the lapse of ages. These progressive stages constitute, in fact, our "systems" and "periods;" and if in one region there should appear to be a sudden break between them, let it ever be remembered that the deficiency is to be supplied by some other district—in other words, let it be remembered that the oscillations of sea and land, of elevation and depression, and other physical changes, are sufficient to account for local breaks in life—but that there is no foundation whatever for the belief in "general extinctions," and, consequently, "new general creations." So far as the few thousand years of man's experience extends, the current era is as mutable as any of the epochs that preceded, and yet so gradually have its extinctions and creations taken place, that science can scarcely corroborate the one, and has as yet failed to detect the other. The systems of the geologist are, therefore, mere concatenations of events indicative of certain periods; and as nature never repeats herself in time, each period, when taken at sufficiently distant intervals, is characterised by *some* forms of vitality peculiar to itself, the while that its *general* life merges imperceptibly into that of the epoch that follows, just as it was imperceptibly interwoven with that which preceded.

[Development Hypotheses.]

This belief in a gradual and unbroken evolution of vitality gives no encouragement to the doctrine of development from lower to higher types, through some long-continued but little-understood process of *physical* transmutation. We say "physical" transmutation; for, whether we appeal, with Lamarck, to the modifying influence of new external conditions—with the author of the *Vestiges*, to the force of internal volition on the embryotic organism—or with Mr Darwin, to the gradual accumulation of minute beneficial changes, which amount in the long-run to specific distinctions, we adopt the same blind-chance process, and are merely phrasing in different terms the same materialistic* hypothesis. Of such a process we have no direct evidence either in existing nature or in that which has become extinct; nor by the assumption of such a process can the various grades and affinities of vitality be logically reduced into one harmonious and consistent scheme. If by any unknown genetic process the polype has given birth to the star-fish, the star-fish to the mollusc, the mollusc to the fish, the fish to the reptile, the reptile to the bird, and the bird to the mammal, it must have been either through a graduated succession of intermediate forms, or at once and directly.

* Should this assertion appear unwarranted, we have only to refer to Lamarck's own avowal, to the advertisement first announcing the publication of the *Vestiges* in 1844, and to the whole tone and tenor of the *Origin of Species*, in which there seems to be a studied non-recognition of any higher Influence than chance, external conditions, nature, law, and other kindred activities.

If by the former process, where is the finely graduated scale of transitional forms, either living or fossil? And if by the latter, how comes it to pass that we have not a universal uniformity of life-type at the successive stages of geological time? for we cannot conceive of any mere physical law acting with discrimination, and peopling one region with one set of forms, and another region with other classes and orders. If it shall be argued that the physical conditions of one region differ from those of another, and must necessarily be accompanied by a diversity of results, we restrict our reasonings to any one area, and there we find as great a complexity and variety characterising the *part* as the advocates of this hypothesis can demand for the *whole*. Some forms continue persistent and unchanged, others die out and are succeeded by closely allied forms; some remain scanty and localised, while others increase and largely extend their boundaries; and this under precisely the same conditions and in the self-same area—a fact altogether inexplicable were the influence of external conditions the only factor in the law of vital diversity.

As to "intermediate or gradational forms," let us take care also that we do not mistake functional resemblance for genetic affinity, and simulative forms for identities. That we have quadrupeds, like the ornithorhyncus, partaking of some of the characteristics of birds; mammals, like the whale, modified so as to assume the aspect of fishes; fish-like reptiles, as the ichthyosaurus; and reptile-like fishes, as the rhizodus, no one for a moment gainsays. These, however, are mere functional resemblances, not genetic affinities; the modification of the great aboriginal types, so as to adapt them for every element—air, earth, and ocean—and to fit them for the performance of every function which the conditions of the world, for the time being, might require. The whale, though swimming in the ocean, is nevertheless a mammal breathing by lungs, bringing forth its young alive, and suckling them with true mammalian affection; and the young-bearing, young-suckling bat, though fluttering through the air like a bird, has no essential feature in common with the birds, save that which belongs to the great vertebrate pattern. Such resemblances are simply *adaptive*, not *essential*. Instead of indicating any genetic affinity, they merely point to a law which ordains that agreement of habit and economy, in widely differing groups, shall be accompanied by similarity of form; and this, of physical necessity, so long as the same element has to be traversed, the same kind of food sought after, and the same general functions to be performed.

Again, if at various stages the lower had given birth to the higher, we should naturally have expected only the lowliest and simplest at first, and an equable and uniform diffusion of the higher races, step by step, in the successive geological epochs. Instead of this, we find protozoans, radiates, articulates, and molluscs, side by side, in the lowest fossiliferous rocks; and in every stage upwards a variety and complexity of higher and lower, which seem to obey anything but a regular arithmetical or geometrical progression, such as any mere physical law of development must necessarily obey. The palæozoic brachiopods were higher and more varied than those of existing waters; the noblest cephalopods—shell-clad and shell-less—were those of the secondary period; the highest structural fishes were the sauroids of the upper palæozoic; and anatomists (Owen) assure us that the thecodont reptiles of the new red sandstone, had they existed at the present day, would have taken rank at the head of the Lacertian order. So far as palæontology can prove, there is no known line of continuous development from one primordial germ—no uniform genetic ascent in time for the various classes and orders of vitality; and the rise and progress which geology unfolds has been clearly under the influence of a much more complicated law than that which takes order from the force of mere external conditions. Besides, in regarding external conditions as the sole cause of vital diversity, we ascribe to them a task to which they are unequal, and leave altogether unexplained which family of the radiates, for example, and why that family alone, was selected to be transformed into the articulates; which of the mollusca became the progenitors of the fishes; or which of the fishes, while the others were left uninfluenced in their piscine state, were advanced to the dignity of giving birth to the reptiles.

If there be any truth in the hypothesis, the highest forms of the lower class must have always given birth to the lowest forms of the next higher class—for we can scarcely expect the lower forms to have been endowed with a power which was not permitted to the higher. Now, so far from this being geologically true, we find fishes making their appearance in the Silurian rocks ages before molluscan life had attained its culminating point in the oolitic era; so also we find reptiles appearing almost simultaneously with fishes in the old red sandstone, and long before the higher fishes (their natural progenitors, according to this theory) had appeared; just as birds are found in the new red sandstone ages before the highest forms of reptiles had come on the stage in the upper secondary series. Here then, the offspring often precedes the parent; and any line of uniform development is altogether disproved by the very facts on which the advocates of the Law of Development are attempting to found it. That during the long lapse of geological time there has been rise within each great subdivision of life from lower to higher forms, no one can deny; but the law of this progress is other than that of mere physical development, and lies, as yet, far beyond the grasp of human philosophy. Nor, indeed, as has been fitly remarked by Professor Agassiz, " will there be any scientific evidence of the method of God's working in nature, until naturalists have shown *that the whole creation is the expression of a thought, and not the product of physical agents.*"

Further, as every creature has its own nature, and habits, and functions, we must, under the transmutation hypothesis, either make the plant and animal capable of changing their own nature and habits, or ascribe the change to the force of external conditions. In the vegetable kingdom we can readily admit, within very wide limits, the operation of external causes; but even there the generic diversity occurring under precisely the same conditions remains altogether inexplicable by such a hypothesis; while the nature of plant-life for ever debars the idea of internal volition. In the animal kingdom we have the same difficulty to contend with. Admitting that physical conditions had the power to modify the vital organism, the nature and limit of these modifications must be predetermined and directed in order to preserve the harmony that prevails throughout living as well as throughout extinct forms; and this harmony can never be other than the ordaining of a governing mind. It is impossible to invest any mere physical law with a discriminating power—absurdity to ascribe to individual volition any permanent change of organisation while an intimate relationship continues to pervade the whole. And even admitting the Creator had chosen to act through such means, they can be placed in no higher light than the unconscious machinery of a system requiring superintendence at every turn, and whose every variation is in effect the equivalent of a new creation. The conversion of a mollusc into a fish, or of a fish into a reptile—even if accomplished by a thousand imperceptible stages—is to our apprehension as much a creative act as the aboriginal formation of the mollusc; and though nature acts largely through the employment of secondary causes, science will ever most safely appeal to the *primal*, till she has learned to determine with precision the operations of the *secondary*—returning, like Noah's dove, from an ocean of inquiry that offers as yet to the sole of her foot no sure and abiding resting-place. No doubt plants and animals are endowed with a certain amount of elasticity so as to adapt themselves to minor changes of external conditions; and acting upon this elasticity, man has been enabled to produce all the varieties of cultivated fruits and grains, and domesticated animals. This limit of variation, however, is soon reached: the *species* is never affected, and the *varieties* can only be maintained by a continuation of the artificial stimulus.* In this case man presents himself as a sub-creative centre, deputed with a power of prescient design otherwise unknown in creation; and to argue from his operations, as Mr Darwin has done, to those occurring in mere physical nature, is altogether to misinterpret the functions that intellect and reason were destined to subserve. As we have no other power in nature akin to the human intellect, so we are not entitled, in the spirit of induction, to argue from the results produced by that intellect to the operations of the unreasoning material agencies of nature.

To appeal, in the next place, to embryology—to state that, in their embryonic stage, the higher animals always pass through the successive phases of those that are lower,

* That the individuals of a species should be capable of varying within certain limits, so as to adapt themselves to minor variations in their creative centres, seems part of a wise and beneficial arrangement; but that such variations partake of a progressive character is disproved rather than supported by the well-known tendency of all artificial varieties to revert to their original stocks.

and then to maintain that the fish-embryo, for example, acted upon by extremely favourable conditions, might be developed beyond itself into a reptile, or the reptile into a bird, is assuming a doctrine of which we have no proof, and merely stating, in perverted terms, the grand physiological fact of animal gradation and affinity. Besides, it is an argument that cuts both ways. If an embryo, under favourable conditions, can be developed beyond its own parent species, it may also, under unfavourable conditions, be retarded and thrown back into the grade that lies beneath it; and as external conditions varied during the geological epochs—now genial and now obnoxious—we ought to have degradation as well as development. The great gradational progress taught by geology being always steadily from higher to higher, is, however, against this; and when a species or family is subjected to obnoxious conditions, it invariably dwarfs and dies out in its own proper character—a trilobite as a trilobite, an ammonite as an ammonite, an ichthyosaur as an ichthyosaurus—and never under the guise of a lower order. All, too, that we know of existing nature is against this doctrine of transmutation—species and genera remaining (under the restricted limits of variation) as fixed and permanent now as they were known to the Ninevites and Egyptians four thousand years ago.

It is argued, no doubt, that the transmutational advances from species to species take place by slow and imperceptible stages, which cumulatively become apparent only after the lapse of ages. Admitting, however, this rate of progress, there ought still to be transitional forms in various stages of progress at every epoch—forms which we fail to perceive in the living world, just as geology has failed to detect them in that which has become extinct. Again, the modifications for which the developist contends are those of a *beneficial* kind; so that, in the great struggle for existence and under the influence of altered conditions, every creature, advantageously modified, will have a chance of surviving, whilst those unaffected must go to the wall. He fails, however, to show how the operation of a purely physical law should not affect alike every member of a species, and to perceive that his doctrine of "natural selection" is but a materialistic phraseology for an undefined law of progress, which forms part of a predestined plan, and must clearly obey an intelligent behest. Above all, he fails to prove how or in what manner it could be more advantageous, in a world where every adaptation is perfect, for a crustacean to drop the mask of a trilobite and assume that of a eurypterite, or for a eurypterite to drop, step by step, its characteristic organisation, and put on the ultimate guise of a lobster. Still further, if there has really been such a perpetual transmutation of form and function, we are driven backward and backward in the abysm of time to simpler and simpler forms, and compelled to seek for herbivorous and carnivorous races a common paternity and origin. To transmute, however, the graminivorous into the carnivorous—to change entirely their every organ of prehension, mastication, and digestion—their habits and instincts and functions—even if it were conceivable, is utterly disproved by the geological record, in which, from the earliest epochs, we find planteater and flesh-eater arranged side by side in the great drama of life, and as sharply defined in all their characteristic organisation as they are at the present moment.

The hypothesis, untenable as it may appear, must be carried still further. As man is inseparably connected with the great scheme of vitality, any genetic doctrine of transmutation must be equally applicable to him as to the rest of creation, and he must stoop, however humiliating, to trace his pedigree from the order that stands next beneath him. Against this, however, all reason and moral instinct recoils. To transmute the monkey into man, even though the change were effected, step by step, through a whole wilderness of monkeys, could not invest the brutish nature with the human intellect, or endow the progeny of the irresponsible beast with the moral responsibility of man! Admitting the similarity of physical organisation—admitting the lowly condition of the lowest varieties of the human race—and granting that the difference between the most highly endowed philosopher and most degraded savage was even greater than that between the lowest savage and the most exalted monkey—still we know of no intermediate forms, living or extinct,* to bridge over the gulf that lies between—no germ of moral perception in the brute, whereon to graft the improving consciousness of moral responsibility in the man. Here then (admitting that men had been *physically* descended from monkeys) there is something in the man unknown and unevidenced in the brute; and unless we can learn to regard this superadded gift of reason and moral perception—to say nothing of religious sentiment—in the light of a new creation, the common ground of argument is removed from between us, and conviction becomes impossible. Even were we to concede the point of mental relationship, and to admit that science could trace the most intimate relations of organic life—that life which associates man with the plants and animals around him; still (as has been

* It has been hinted, no doubt, that as other mammalia have had their gigantic tertiary precursors, so it is likely the gorilla and chimpanzee were also preceded by larger and more man-like forms of monkey. Of the existence of such forms we have not, at present, the slightest indication; but, admitting the ingenuity of the surmise, and supposing such remains were to be discovered to-morrow, it will still remain to be shown that larger and more erect aspects of ape must necessarily be endowed with higher mental and more man-like qualities.

aptly remarked) " no observation from the outside ever did, or ever will, approach that most intense of all realities— our relations as responsible agents to right and wrong." This is the rock ahead on which all theories of mere physical development must ever split; and their abettors are driven to this dilemma—either to maintain the identity of man's nature (though differing in degree) with that of the beasts that perish, or frankly to admit that the human race sprang into being only when "God breathed into his nostrils the breath of life, and man became a living soul."

We are aware that certain physiologists who adopt the development hypothesis contend also for a unity of mental constitution between man and the lower creatures — its manifestations differing only in degree among the various grades of organisation. Adhering to the one hypothesis, they are prepared to accept the other and all its consequences as a logical and sequential deduction. It is strange, however, to find others who, like Professor Agassiz, repudiate all theories of physical development, adopting a similar conclusion; and not only so, but arguing for the community of an immaterial and immortal principle, as if this were not a stronger argument for universal genetic connection than any that can be drawn from mere similarity of external organs. "For the most part," says the Professor, in his *Essay on Classification*, " the relations of individuals to individuals are unquestionably of an organic nature, and, as such, have to be viewed in the same light as any other structural feature; but there is much also in these connections that partakes of a psychological character, taking this expression in the widest sense of the word. When animals fight with one another—when they associate for a common purpose—when they warn one another in danger—when they come to the rescue of one another— when they display pain and joy—they manifest impulses of the same kind as are considered among the moral attributes of man. The range of their passions is even as extensive as that of the human mind, and I am at a loss to perceive a difference of kind between them, however much they may differ in degree and in the manner in which they are expressed. The gradations of the moral faculties among the higher animals and man are moreover so imperceptible, that to deny to the first a certain sense of responsibility and consciousness, would certainly be an exaggeration of the differences which distinguish animals and man. There exists, besides, as much individuality, within their respective capabilities, among animals as among man, as every sportsman, every keeper of manageries, and every farmer or shepherd can testify, or any one who has had large experience with wild, tamed, or domesticated animals. This argues strongly in favour of the existence in every animal of an immaterial principle similar to that which, by its excellence and superior endowments, places man so much above animals. Yet the principle unquestionably exists, and whether it be called soul, reason, or instinct, it presents in the whole range of organised beings a series of phenomena closely linked together; and upon it are based not only the higher manifestations of the mind, but the very permanence of the specific differences which characterise every organism. Most of the arguments of philosophy in favour of the immortality of man apply equally to the permanency of this principle in other living beings. May I not add, that a future life, in which man would be deprived of that great source of enjoyment and intellectual and moral improvement which result from the contemplation of the harmonies of an organic world, would involve a lamentable loss? And may we not look to a spiritual concert of the combined worlds and all their inhabitants in presence of their Creator, as the highest conception of paradise ?" For hypotheses such as these, however curious or startling they may appear, let it be clearly understood that science is in no way responsible. No observation from the external world—no analogy, however plausible—no analysis, however minute—can ever solve the problem of an immaterial and immortal existence. They may be received as possible or probable auxiliaries, but in the main our faith on this point must rest, as it has hitherto rested, on an altogether different foundation. Science has its own line and limit of inquiry, and no satisfactory result can ever arise from any attempt to carry it beyond the boundary of the philosophically attainable. If the developists have failed on physical grounds to prove a genetic unity for the various grades of organisation, their opponents only complicate the question by the unnecessary introduction of the still more difficult problem of a spiritual community.

[Acceptance of Vital Hypotheses.]

While repudiating this doctrine of physical development, we would treat its advocacy without that acrimony and invective which has been too frequently displayed against it. The progress and gradation of vitality is still in a great measure a mystery to science; and any honest and earnest endeavour to unveil it should ever meet with a corresponding regard. In the organic as in the inorganic world the Creator often operates through secondary causes, and the discovery of these causes, in the spirit of true philosophy, is to human reason a duty as well as a privilege. Every result that meets the senses, every phenomenon that nature presents to us, becomes the legitimate subject of scientific research; and subtle as the eliminations of Life may be, mysterious as its ordainings may appear, there is clearly nothing in its character to put it beyond the pale of such investigation. Where, then, so little is positively known, and so much merely tentative and temporary, no one has a right to dogmatise*—far less to treat the earnest opinion of another otherwise than in the spirit of candour and respect. Argument is weak if it cannot divest itself of acrimony; truth is half shorn of her lustre when surrounded by a medium of angry invective. The development hypothesis, when pursued in a right spirit—in the spirit of inductive research and logical interpretation—is entitled to a fair hearing, even should it startle our accustomed beliefs and offend our prejudices. Science, confident in its strength, grapples with the argument; prejudice, feeling her weakness, avoids the combat, and, assassin-like, launches those infernal missiles—" sceptic," " infidel," and " atheist." But whatever the uneasy tenderness with which the theme of Life is usually treated, its origin and progress, its incomings and outgoings, are questions which meet us at every turn in geology, and themes which no scientific naturalist can possibly ignore. Year after year they are being more forcibly pressed upon our attention, and no geologist can afford to stand by while the brunt of the battle must be met on the ground of his own special science. Lamarck's well-known hypothesis—the *Vestiges of Creation*, which stands bastardised by the moral cowardice that shrinks from avowing its paternity—and Mr Darwin's *Origin of Species*—have each given a fresh impetus to the question; and though our limits debar any further discussion of the question, we may be permitted to express our opinion, that be it " trans-

* " In respect to very many questions, a wise man's mind rests long in a state neither of belief nor of unbelief. But your intellectually shortsighted people are apt to be preternaturally clear-sighted, and to find their way very plainly to positive conclusions upon one side or the other of every mooted question."—Dr ASA GRAY, in his *Review of the Darwinian Hypothesis*.

mutation under the influence of external conditions"—"development through the force of maternal volition on the embryotic organism"—or, "natural selection in the struggle for existence," neither of them (even were they true to the extent their advocates argue) ascends any higher than a mere subordinate factor in the law of vital development. We are far from denying the influence of such causes on the diversity of life. On the contrary, unprejudiced inquiry is constrained to rank them among the activities of the Creator's plan, but simply as secondary activities, limited alike in their power and in the range of their applicability. Thus, however, it ever is: we discover a cause where several others are equally operative and potent, and our ignorance or enthusiasm is but too prone to ascribe to the one what is ascribable alike to the others that remain undetected and undetermined.

Even Mr Darwin, wedded as he is to the theory of Natural Selection, is constrained to admit the operation of several activities in the law of vital diversity. " It is interesting," he says, in one of the most genial passages in his work, "to contemplate an entangled bank, clothed with many plants of many kinds, with birds singing on the bushes, with various insects flitting about, and with worms crawling through the damp earth, and to reflect that those elaborately constructed forms, so different from each other, and dependent on each other in so complex a manner, have all been produced by laws acting around us. These laws, taken in the largest sense, being growth by reproduction ; inheritance, which is almost implied by reproduction ; variability from the indirect and direct action of the external conditions of life, and from use and disuse ; a ratio of increase so high as to lead to a struggle for life, and as a consequence to natural selection, entailing divergence of character and the extinction of less improved forms. Thus, from the war of nature, from famine and death, the most exalted object we are capable of conceiving—namely, the production of the higher animals—directly follows. There is grandeur in this view of life, with its several powers having been originally breathed into a few forms or into one; and that, whilst this planet has gone cycling on according to the fixed law of gravity, from so simple a beginning, endless forms most beautiful and most wonderful have been, and are being evolved." Here then, according to his own showing, inheritance, external conditions, use and disuse, struggle for life, and natural selection, are all fulfilling their parts as co-factors in one great law, and it is strange that in the face of this admission he should labour to ascribe to one cause what would have been much more philosophically and satisfactorily ascribed to the many. He admits, too, the " original breathing of life into a few forms or into one form," and yet unaccountably appeals throughout his argument to *chance* and *nature* for all subsequent development, as if these blind deities were aught without the direction of the same original life-breathing Impulse! If science is constrained to admit a Divine origination of life, why should she be ashamed to confess to an equally Divine sustaining of its subsequent manifestations? If we are compelled to invoke a creative act for a beginning we cannot comprehend, why should we shrink from appealing to the same cause for subsequent diversities we cannot explain? But for this weakness or vanity, the erroneous in these so-called " theories of life" had met with a kindlier tolerance, and the true with a readier acceptance.

If, as these theorists assert, the question be merely this: Has or has not the Creator endowed inorganic matter with the power of assuming, under the influence of certain forces, an organic form? and has or has not the Creator further ordained that under certain external phases of nature these forms shall be transmuted into other and altered forms of organisation? then the subject assumes a purely physical aspect, and they are bound, like the mathematician and chemist, to prove their case by the ordinary rules of physical induction. Given the scales, fins, and gills of a fish—what the conditions and what the amount of time necessary to transmute them into the scutes, paddles, and lungs of a marine reptile? Given the scutes, membranous fore-arms, and stomach of a flying reptile—what the phases of change and what the amount of time required for their transformation into the feathers, wings, and gizzard of a bird? Or, given the four hands with partially opposable thumbs, the low facial angle, and the jabbering half-reasoning instinct of a monkey—what the force of conditions, and what the term of time for their development into the two-handed dexterity, the erect aspect, and the eloquent ratiocinations of a philosopher of the nineteenth century? If the question be one of purely physical import, such are the formulæ the developists are called upon to frame, and such are the problems that await their solution. This task they have hitherto failed to accomplish; and as yet the place of sterling proof is usurped by plausible assumption. The evolution of life, however, in all its multifarious forms and aspects—its cosmical functions and relationships—its orderly appearings and disappearings at certain geological periods—its bearings on the intellectual and moral position of Man—all this and much more that instinctively interweaves itself with our innermost thoughts of time and destiny, must surely rest on a broader and deeper foundation. It is—if anything we shall ever comprehend—the gradual unfolding of a predestined plan, the expression of a Divine thought, which it is our high privilege as well as duty to interpret; but depend on it, we altogether err in our method of interpretation if we attempt to associate life with physical agency in any other way than the mere medium through which creative power has chosen to manifest itself to our observation. In vain does Mr Darwin taunt that this is a mere "dignified way" of putting the question: better surely to rest satisfied with a dignified belief we are unable to prove, than seek unsatisfactory shelter under a cold undignified materialistic assumption! For our own part, believing as we do that Life in all its relations—its incomings and outgoings in time—its modifications in form, and its distribution over space—are under the incessant operation of fixed and determinable laws, we are as free to entertain the question of vitality as we are to entertain the formation of a stratum of sandstone or the aggregation of a mineral crystal; but this we cannot do unless at every stage of our reasoning we associate a superintending with a creative intellect. And we have yet to learn wherein the variation of a natural law, or the variation of a well-known form of life—even to the ten-thousandth degree—is less an act of creation than the original establishment of that law, or the original calling of that life-form into existence.

[Advent of Man.]

The study of life, palæontologically regarded, necessarily involves the creation and first appearance of Man; and on this subject much discussion has taken place, unprofitable alike to science and the cause of Christian theology. So far as geological evidence goes, we have no traces of man or of his works till we arrive at the Superficial Accumulations—the coral-conglomerates, the bone-breccias, the cave-deposits, and the peat-mosses of the current period. It is true, that so far as the earlier formations are concerned, the evidence is purely negative; but taking into account all that palæontology has revealed touching the other families of animated nature, the fair presumption is, that man was not called into being till the commencement of the current geological era, and about the time when, in the northern hemisphere, the sea and land received their present configuration, and were peopled by those genera and species which (with a few local removals and still fewer extinctions) yet adorn their forests and inhabit their lands and waters.

It has been often argued, that up till this time the world was altogether unfit for the habitation and support of Man —its physical conditions being so unstable, and its flora and fauna being unsuited for his sustenance. Now, while we at once admit a physical as well as a moral fitness in all things created, and that no creature was brought on the stage of being till external conditions were suited alike for the maintenance and genial development of its existence, we must guard against any hasty generalisation that is not absolutely warranted by the facts of geology, and which, in its ultimate bearings, is quite as materialistic and physical as any other that has been advanced to account for the phenomena of vital development. The idea of a generally unstable and convulsed world during the earlier geological epochs is altogether disproved by the facts we have over and over again repeated, even if it were not abhorrent to all philosophical notions of a law-regulated cosmos; and the alleged absence of plants and animals necessary for man's sustenance scarcely rests on a surer foundation. It is true, and a beautiful corroboration of the fitness of physical conditions, that all the flowers, and fruits, and cereals, all the domesticated animals—the horse, ox, and sheep—on which man in temperate regions so much relies for the comforts and necessaries of existence, are unknown till the latest geological epochs — the means of support occurring simultaneously with the object to be supported. But while this holds true, and is fitly applicable to a beef-cooking, bread-eating phase of human progress, it is not strictly applicable to man in all his conditions; and it is quite conceivable (geologically speaking) that inferior races of men may have existed in much earlier epochs. The flora and fauna of the oolite are extremely similar to those of Australia, where we know that an early aboriginal race have for ages hunted in the bush and camped on its grassy karoos. The inhabitants of the South Sea Islands live exclusively on palm-fruits, on farinaceous roots, and the fish of the surrounding ocean: now, palm-fruits and farinaceous roots abound in the lower tertiary and in the oolite, and we see nothing in the fishes of those periods that would render them inedible or unnutritious. The Esquimaux, to whom the very names of *tree* and *wheat* are unknown, and who exist on fish, seal-oil, and whale-blubber, among the extreme rigours of the north, attain even there a certain amount of civilisation; and such a lowly race would have found precisely similar conditions in Middle Europe during the glacial era, when icebergs floated in our seas, and whales and seals were stranded in our estuaries.

We mention these things not from a conviction that man existed during those early epochs, but simply as an argument to show that his first appearance, at whatever period, must have been in accordance with the general plan of vital development, and not in obedience to any phase of external conditions, and that we may fairly expect, in the progress of geological discovery, a much higher antiquity to be proved to the human race than is now usually assigned it. And even now, proofs are not wanting in the lake-deposits, the bone-caves, and peat-mosses of Southern Europe, to connect man with the latest pliocene fauna, and to render it possible that he contested the same cavern with the lion and hyæna, hunted the gigantic Irish deer on the plains, and speared the mammoth and mastodon in its forests. Nor would such a discovery militate in any way against the facts of history, so far as these are known, with anything like demonstrable certainty. The facts and their order remain the same; it is only the chronology, about which the ablest historians still differ so widely, that could possibly be affected. Moreover, while reasoning about the advent and progress of man, let it ever be remembered, that the higher the race the more rapid its culmination; and that we have no standard in the development of the lower creatures impelled solely by instinct, wherewith to measure the progress of man guided by reason, and capable of making the elements of nature subservient to his elevation and to his dispersion over every region of the globe. On the whole, and as geological evidence now stands, man, though the noblest, is one of the latest emanations of creative wisdom—crowning, as it were, that long line of gradational vitality which apparently began with the Silurian epoch, but whose further progress and termination lies in the mysteries of the future.

And here it must be observed that Geology, though often indiscreetly summoned to pronounce, can throw little or no light on certain questions respecting the advent and early condition of our race. The varieties of the human family, distinguished and described by ethnographers, are altogether unknown to Geology; and, so far as the stone-implements, the cave-fires, and the tree-canoes of the pleistocene epoch are concerned, they are such as might readily be formed by any of the savage tribes of the present day. Whether, therefore, man originated in one centre or in many centres—whether the several known races are separate creations, or merely time-distributed varieties, of the same one-created species—Geology can give no certain reply. Not a skeleton or skull has yet accompanied these primeval implements to indicate the character of the race that fashioned them; and be it ever remembered that only the merest specks in Western Europe have yet been examined—leaving wholly untouched the wider areas of Asia, to which history and tradition alike point as the earlier nursery of the human family. In the mean time then, these ancient implements, wherever they occur, indicate the same conception and the same design, and would go to prove—so far as the evidence is of value—a unity and community of the heads and hands concerned in their fabrication. Another question occasionally mooted by theologians who dabble in geology, is, that man came from the hand of his Maker a higher and nobler being than those rude old implements would seem to imply, and, therefore, they are of no antiquity, but the mere yesterday fabrications of a savage and curse-degenerated race. Geology, restricting herself to her own proper province, declines to argue this question. Those are the facts deep in the old alluvia and gravels; these are the evidences which science has to deal with; and sound induction will not permit her to travel beyond her own tangible record. The question of moral debasement lies altogether beyond the pale of natural science; and these disputants seem to forget that man's early condition, as indicated by geology, is much the same as that depicted in the Mosaic record. Our great progenitors sewed themselves aprons of fig-leaves, not of jacquarded silks or power-loom calicoes; were tillers of the ground and keepers of flocks, ignorant alike of high farming, steam-ploughs, and reaping-machines; travelled and communicated by camel-caravans and pack-horses, not by railroads and electric telegraphs. The course of civilisation is ever slow and gradual; and history, tradition, and experience alike point to the early condition of every race as hunters and herdsmen—the conditions unmistakably indicated by those simple pleistocene implements.

[Time Geological.]

Whatever may have been the creational development of plants and animals—whenever the advent or whatever the first condition of the human race—the groups and systems of geology afford irrefragable evidence of the lapse of vast epochs of TIME. The idea of immense duration is at once suggested by an examination of the stratified rocks. The innumerable alternations of their shales, limestones, sandstones, and conglomerates—their vast thickness—their repeated laminations—the alternation of marine and freshwater beds—their upheaval into dry land and subsequent submergence, again and again—the various races that have lived and grown and been entombed in them, system after system—all this, and much more that will readily suggest itself to the reflecting mind, must clench beyond cavil the conviction of the unconceivable duration of geological time. In all our reasonings, then, we must never lose sight of the element TIME. With unlimited duration at command, we have a power equal to the mightiest results; and forces which in themselves appear puny and feeble, become giants when backed by that spirit of unrest whose eye never closes, whose wing never wearies, and whose foot never tires. The hardest rock is hollowed by the ceaseless waterdrop; the Nilotic plain has been borne, particle by particle, from the mountains of Abyssinia; and the massive coralreef of a thousand leagues owes its origin to an animated speck all but invisible to the unassisted eye. The problems of geology, like the problems of mechanics, are thus dependent for their solution on the conjoint elements of *force* and *time*. Where the exertion of force is great, the time for performance may be short; but where the force is small, the time must be proportionally prolonged. The two elements are ever in inverse ratio; and thus agents in themselves comparatively insignificant may, during the lapse of ages, accomplish most important results. It is generally, therefore, to the cumulative effects of this inexhaustible resource of TIME that the developists make their last appeal—contending that the progress of transmutation is so gradual as not to be appreciable with the five or six thousand years of man's observation. Admitting the plausibility of the argument in existing nature, the geologist appeals to the fossil world for evidence of these insensible gradations, and he finds stratum after stratum containing the same unchanged species, and then, in the next stratum, at once and decidedly, the remains of a species altogether new. This, if anything we can intelligibly define, is not genetic gradation, but pre-appointed creation; and as time is merely passive unless the law of specific progress obeys some active and controlling power, an eternity of time will never affect it.

In the present state of our knowledge, any attempt to calculate geological knowledge by years is altogether futile; we can only indicate its vastness by the use of indefinite terms, as "eras," and "epochs," and "cycles." It is customary, however, to speak of *pre-geological, geological*, and *historical* time—meaning by pre-geological all that extends backwards before the deposition of the fossiliferous rocks; by geological, all that is embraced between the earliest fossiliferous deposits and human history; and by historical, all to which a determinate chronological value can be assigned. With regard to the first, it is an abysm which the human intellect, even in its boldest flights, shrinks from exploring; as to the last, important as it may seem to man, creationally it is but a thing of yesterday; while to time geological we turn as entering into every problem of our science, and investing their consideration with strange and deeper interest. The amount of this time we have as yet no means of estimating—no power to give it expression in years and centuries. Many ingenious calculations have, no doubt, been made to approximate the dates of certain geological events, but these, it must be confessed, are more amusing than instructive. For example, so many lines of mud are annually laid down by the inundation of the Nile, fragments of pottery have been found at the depth of thirty feet. How many years have elapsed since the pottery was first imbedded? Again, the ledges of Niagara are wasting at the rate of so many feet per century. How many years must the river have taken to cut its way back from Queenstown to the present Falls? Again, lavas and melted basalts cool, according to the size of the mass, at the rate of so many degrees in a given time. How many millions of years must have elapsed (supposing an original igneous condition of the earth) before its crust had attained a state of solidity? or, farther, before its surface had cooled down to the present mean temperature? For these and similar computations it will at once be perceived that we want the necessary uniformity of factor; and until we can bring elements of calculation as exact as those of astronomy to bear on geological chronology, it will be better to regard our "eras," and "epochs," and "cycles" as so many terms, indefinite in their duration, but sufficient for the magnitude of the operations embraced within their limits. But even on this point of expressible time, the earnest geologist is not without hope and encouragement. He rests confident (confident as in the existence of his own being) that the whole history of geological phenomena—the shifting of volcanic energy from centre to centre, the elevation and depression of certain areas of the earth's crust, the interchanges of sea and land thereby occasioned, the recurrence of colder and warmer climates over determinable latitudes, the necessary re-arrangements of life attending these changes, and the like—is but a chronological exposition of the influence of natural law; and that as law is as obedient to *times* as to *modes*, the periodicity of these occurrences will one day or other be determined. This done, its expression in years and centuries is a simple task; but though accomplished to-morrow, and expressed in figures like the distances of the astronomer, the mind would altogether fail to grasp the conception of its immensity.

[Course of Creation.]

On the whole, then, the systems and cycles of the geologist—imperfectly interpreted, as they yet undoubtedly are—present a long series of vital gradation and progress. Not progress from imperfection to perfection of purpose, but from humbler to more highly-organised orders, as if the great design of Nature had been to ascend from the simpler conception of *materialism* to the higher aims of mechanical combination, from *mechanism* to the subtler elimination of mind, and from *mentalism* to the still higher attribute of *moralism* as developed alone in the heart and soul of man. Thus, from a long azoic period, during which the material elements of the world were being eliminated into mechanical order under the influence of chemical and physical forces, we rise, as it were, to the conception and first expression of vitality in the simple organisms of Cambria and Siluria. Again, from the lowly sea-weeds of the silurian strata and the marsh-plants of the old red sandstone, we rise (speaking in general terms) to the prolific club-mosses, ferns, reeds, and gigantic endogens of the coal-measures; from these to the palms, cycads, and pines of the oolite; and from these, again, to the exogens and true timber-trees of the tertiary and current eras. So also in the animal kingdom: the graptolites and trilobites of the silurian seas are succeeded by the eurypterites and bone-clad fishes of the old red sandstone; these by the sauroid fishes of the coal-measures; the sauroid fishes by the saurians and birds of the trias and oolite; the reptiles and marsupials of the oolite by the true mammals of the tertiary epoch; and these, in turn, give place to existing species, with man as the crowning form of created existence. And even as regards man, he, too, has ever been in a state of gradation and progress. Many ancient races and forms of civilisation have passed away, and others have taken their place. Nor has the line of development in man's case been uniform and continuous, any more than in the purely geological elimination of vitality. Here at one time, and there at another, with greater intensity—now torpid and slow, now fresh and vigorous, but ever and always still forward—the human mind acquiring a cumulative force from the experience of the past, and that race becoming most powerful who can grasp the most of nature's laws, and turn them with irresistible force to its own purposes. As in the great design of Nature, so in the minor scheme of Humanity that lies within it, the progress has ever been from materialism to mechanism, from mechanism to mentalism, and all that science indicates or history reveals points to moralism as the highest stage of man's terrestrial development.

Such is clearly the course of creation, however dimly we may descry the law that governs its elimination. Matter acted on by certain forces assumes the varied mechanism of minerals, plants, and animals; to this graduated mechanism is gradually superadded the qualities of sensation and mentalism; and to mentalism in its highest phase is bequeathed the godlike gift of moral perception. Much of the similarity that runs through the great types of Life has evident reference to those physical forces which act independently alike on all matter, organic and inorganic; but over and above this, there is the homology of parts in the several main divisions of plants and animals—the embryonic phases of life which harmonise in a wonderful manner with the successive geological phases—the ascent in time as well as in organisation from acrogens to endogens, gymnogens, and exogens, from cold-blooded water-breathers to cold-blooded air-breathers, warm-blooded water-breathers, and warm-blooded air-breathers—the curious modifications in time on the various families of the same great classes as already indicated in the geological record—the occurrence of contemporary representative species in distant geographical areas—the similarity of form accompanying the similarity in function in widely separated classes, &c.—all of which are undoubtedly the results of some great pre-appointed and continuously operating law. It may not be the force of external conditions, the power of hereditary impulse affecting embryonic germs, the result of natural selection in the struggle for existence, or any one of the causes acting gradually through indefinite time, that have been advanced to account for the phenomena. Yet each and all of them may be factors in some great scheme of causation; and we are bound in the spirit of true research not only to treat fairly, but to honour, every earnest endeavour towards the solution of the problem. It has been well and prettily said, that "before common minds can know, men of genius must guess; and if in assaulting the citadel of the unknown they should sometimes fall, their names ought at least to be chronicled with honour." In this spirit, and as tentative aims at Truth, suppositions cannot be debarred from our science, and all the less in questions so intricate and obscure as the origin and progress of Life. It may be that the solution of the problem lies far, as yet, beyond the efforts of inductive science, but assuredly the time will come for its attainment, and all the more quickly the less we attempt to dissociate from nature's operations the ever-active presence of a Supreme Intelligence.

In the mean time, all that can be asserted under the warrant of geology is, that in the Vegetable World the course of creation has evidently been from the amphigens of Siluria to the acrogens of Devonia and the coal-measures —from the acrogens to the gymnogens of the coal and new red sandstone—from the gymnogens to the endogens of the oolite—and from these to the exogens of the tertiary and current epochs. This ascent in time harmonises in the main with advance in structural organisation; and, were the geological record perfect, it is more than likely that intermediate forms—like the sigillariæ and lepidodendra of the coal-measures—might be found throughout, linking these great sections more intimately into one continuous series than inosculating species now connect and bind together existing genera. As we approach the present day, the structural forms become higher and more complex; and as we descend in time, step by step the higher disappear,—till ultimately, in the lowest fossiliferous rocks, we meet only with the cellular amphigens, that take rank at the bottom of the botanical scale. In the same way with the Animal World, we clearly ascend from the radiates and articulates of Cambria to the mollusca of Siluria—from these to the fishes of upper Siluria and Devonia—from these to the lowly reptiles of the carboniferous—from the reptiles to the birds of the trias—from the birds to the marsupials of the oolite—and from these to the true mammalia of the tertiary and current eras. Here is the same chronological and physiological harmony: and not only so, but within each great section there has been a similar structural ascent, an ascent (take the crustacea for example) from trilobites to eurypterites—from eurypterites to limuloid forms—from these to the long-tailed lobsters and cray-fish (macrura)—and from the macrura to the short-tailed crabs (brachyura) of the chalk and tertiary—the ancient forms being characterised in their mature state by certain features which now only transitorily appear in the embryonic stages of their existing congeners. In the main, the chronological and physiological harmony is complete; and were the record entire, a thousand connecting forms would appear, linking the whole into one continuous unity of design—a design to whose perfection every part conspires, and yet maintains its own essential and distinctive character. Whatever, we again repeat, may be the operating causes in this scheme of vital evolution, it is clearly the predestined scheme of a Governing Mind—a mind that from the beginning has co-adapted and co-adjusted all the forces and progressive conditions of the universe, and whose power, wisdom, and goodness are the same, whether displayed in a succession of creative acts which we cannot comprehend, or in a series of secondary causations which we fail to explain. If the course of creation be the result of a succession of creative acts, these acts have always the most intimate relation to one another, as well as to those that have preceded, and their order and character are therefore hopefully determinable; if, on the other hand, the course of creation depends on a series of secondary causations, these causes, being patent to our investigation, must be inductively discoverable. Either way, the constitution of our intellect constrains us to inquire; and though the problem may never be fully solved, the effect of the inquiry must be to elevate the creature who earnestly strives to attain to the comprehension of the designs of its Creator.

[Creation still in Progress.]

This idea of progression implies not only an onward change among the rock-materials of the earth in obedience to the physical laws of the universe, but also, as plants and animals are adapted to, as well as influenced by, external conditions, the creation of new species and the dropping out of others from the great scheme of animated nature. And such, we have seen, was the fact even with respect to the current era. The mastodon, mammoth, and other huge pachyderms that lived from the tertiary into the modern epoch, have long since become extinct, leaving their bones in the silts and sands of our valleys. The Irish-deer, urus, bear, wild-boar, wolf, and beaver, are now extinct in Britain; and what takes place in insular districts must also occur, though more slowly, in continental regions. The dodo of the Mauritius, the æpiornis of Madagascar, and the dinornis of New Zealand, are now matters of history; and the same causes that led to the extinction of these, are hurrying forward to the obliteration of the beaver, apteryx, ostrich, elephant, kangaroo, ornithorhynchus, and other animals, whose circumscribed provinces are gradually being broken in upon by new conditions. And here the question naturally occurs, If we have now local removals and general extinctions, what of New Creations? The local removal or the general extinction of any well-known creature we readily perceive; the introduction of new species (unless we assume with Mr Darwin that all varieties are but incipient species) has as yet escaped detection, or resolved itself into that more facile solution—"the discovery of a new plant or animal." Unless, however, creative energy be waxing faint, and the scheme of vitality be destined to come to an end, new creations must take place as infallibly as extinctions. We rest on this as a matter of faith, though human observation has hitherto been so partial and limited, that it is only of late it has been enabled to establish the one, and is just beginning (in the question of the variation of species) to direct attention to the other.

There is, however, no error more common than to consider creation as a thing accomplished—to regard it as an *act* rather than a *work* still in progress. Go back to the earliest condition imagination is able to picture—condense this globe from nebulous masses floating in ether—cool and consolidate its crust from igneous matter, or call it at once into being by the fiat of a word, it is *now* as it was *then* a scheme in the process of creation. All its rock-matter has been and is continually changing, and assuming new forms and distributions. The muds and sands of our present shores will be the rocks of some future hills, and the rocks of our hills the sediments of the ocean of some after epoch. So in like manner with its Vitality. The genera and species have been continually changing and pressing forward under the operation of pre-appointed laws to new and different forms. Call this by what name you will, it is in purpose as it is in effect, *Creation*. Interpose a thousand secondary causes—establish a law for every act, and try to remove by the widest distance the worker from his work—still, these "causes" and "laws" are of themselves utterly impotent, unless sustained and directed as immediately now as they were when first "the Spirit of God moved upon the face of the waters." As we fail to detect any symptom of decay, so we cannot admit the idea of cessation, but must believe in the advent of new races as implicitly as we believe in the physical changes which more directly and forcibly appeal to our observation.

[Duration of Species.]

In reasoning on the causes which have led to the extinction of races, we must not lose sight of the speculation, that species, like individuals, may have had a limit of duration assigned to them from the beginning, and that this limit may be attained even when all extraneous causes remain quiescent and stationary. "Attempts have been made," says Professor Owen, "to account for the extinction of the race of northern elephants (the mammoth of Siberia) by alterations in the climate of their hemisphere, or by violent geological catastrophes, and the like extraneous physical causes. When we seek to apply the same hypothesis to explain the apparently contemporaneous extinction of the gigantic leaf-eating megatherium of South America, the geological phenomena of that continent appear to negative the occurrence of such destructive changes. Our comparatively brief experience of the progress and duration of species within the historical period is surely insufficient to justify, in every case of extinction, the verdict of violent death. With regard to many of the larger mammalia, especially those that have passed away from the American and Australian continents, the absence of sufficient signs of extrinsic extirpating change or convulsion makes it almost as reasonable to speculate with Brocchi on the possibility that species, like individuals, may have had the cause of their death inherent in their original constitution, independently of changes in the external world; and that the term of their existence, or the period of exhaustion of the prolific force, may have been ordained from the commencement of each species." We can readily account for the annihilation of races by the submergence and elevation of land, by alterations in the aërial and oceanic currents which affect the temperature of a region, or by the destruction of their food through climatic changes; but when races wane and die out without any apparent change in external conditions (just as individuals appear, grow up to maturity, and then fade away), we are driven to some such conclusion as the limited duration of specific force. And if species thus depart without the operation of physical causes, we are compelled to accept the converse, that they may also make their appearance independently of the influences of those external conditions on which the Transmutationists have based so much of their hypothesis.

Nor is it the narrow circle of species alone, but the larger groups and families seem also to have had a similar limit assigned to their duration. The graptolites of siluria, the palæozoic trilobites and eurypterites, the carboniferous sigillariæ and lepidodendra, the ammonites of the oolite, the enaliosaurs and dinosaurs of the same epoch, and the palæotheres of the tertiary, all have had their beginning, their culmination, in individual bulk and specific variety, their declension and decay; and this, be it observed, under no phases of external conditions that geology can determine, but apparently in obedience to some law of structural evolution which runs its course within a definite period. The whole system of life, vegetable and animal, appears but to be a pre-arranged series of typical ideas, each to be realised at a certain period and within certain limits of variation, and when once realised to become passive for ever. The realisation of these creative ideas must of course be accompanied by a thousand co-relative circumstances, and the great caution of philosophy should be to avoid confounding concomitants with causes, or mistaking mere ordinal arrangement for sequential connection.

[Term of the Human Race.]

This curious speculation as to the inherent limit of species suggests another equally curious, and of still greater import to man. Generally speaking, the species that has the widest geographical range has also the longest duration in point of time—this wider range increasing its chances of surviving the occurrence of local catastrophes and the vicissitudes of climate. Man, of all animals, is the most cosmopolitan in his nature, being found, in one or other variety of his species, in every region of the globe. It might, therefore, be naturally inferred from this that the existence of the human race will be of corresponding duration; and this inference, geologically speaking, would be correct were it not for another law that seems to regulate vitality. Throughout the whole systems of geology the higher seems to have a more limited duration than the lower orders, their persistence in time being inversely proportional to their biological pre-eminence. Thus, the mammalia of the tertiary epoch had a briefer existence than the reptiles of the Wealden and oolite; these reptiles a more restricted time-range than the palæozoic crustacea; and these again a more limited base of specific duration than the lower shell-fish and corals, some of which in their generic aspects (like the *lingula* and *terebratula*) are extant to the present day. If, then, this generalisation can be established, we are forced to the conclusion that the term of the human species, as well as of those domesticated animals on which he so much relies, notwithstanding their wide geographical range, will be brief in proportion to their organisation.

Startling as such speculations may be, it must ever be remembered that there are a thousand qualifying circumstances in Man's case which do not apply to our reasonings regarding the lower animals. He has not only a wider geographical range, but is more omnivorous in his diet than other creatures, the while that his structure and intellect enable him to combat with difficulties of food and climate to which they must at once succumb. Submerge Australia, and you at once annihilate almost the entire division of marsupial mammals, while as regards the aboriginal tribes you destroy only a very small, and that by no means an important, section of the human family. Alter the climate of Europe so as to place it beyond the growth of the fruits and grains on which its people now chiefly subsist, and, while its high civilisation might be destroyed, and its dense population reduced to a few nomadic races struggling with the rigours of the new climate, still you do not annihilate *them*, nor greatly affect the inhabitants of the other quarters of the globe. Direct the Gulf-stream from its present course across the Atlantic, and you at once destroy whole families of marine life dependent on its thermal waters, and whole phases of vegetation that border the shores where its climatic influence impinges ; but, so far as man is concerned, you only make him shift ground and seek in other localities the congenial conditions that have been withdrawn. Unless, therefore, the human species has a limited creational term assigned to it in accordance with that law which the more highly organised existences seem to obey, we at once perceive that the adaptability of man's constitution, and the inventive powers of his intellect, give him, geologically speaking, every chance of a prolonged tenure of this earthly domain. It must be borne in mind that speculations such as these refer to man only in his geological bearings, and touch not at all on the rise, progress, and decay of nationalities, or types of civilisation. These obey an altogether different set of laws, having reference to that Moralism which separates him from other living beings, and confers on him his highest and most distinctive characteristics.

[Influence of Man on the Future.]

The removal and extinction of species, viewed in connection with the physical changes that are continually taking place on the surface of the globe, necessarily lead to speculations as to the conditions and phases of the FUTURE. Respecting these, however, it seems vain to offer even the widest conjecture, so long as we remain in ignorance of the law that has regulated the progression of the Past. Where the terms of a law are known, the formulæ may be readily framed for the calculation of its times and results; but where these terms are little better than guessed at, our reasonings can never rise beyond the value of the merest hypotheses. Subjected as our planet is to numerous modifying causes, we know, however, that vast changes are ever in progress, and that the present aspects of nature will not be the same as those she must assume in the eras that are to follow. But what may be the nature and amount of these changes, what the new conditions brought about by them, or what the races of plants and animals adapted to these conditions, science has yet no available means of determining. And yet, as we have seen that in past ages certain species of one epoch always passed less or more numerously into the succeeding epoch, so it is not unreasonable to presume that many of the existing species will pass into the period which is to follow. We have also seen that though in certain regions extinctions took place rapidly and entirely, yet over the whole world the progress of vitality has been gradual and continuous; and, generalising in like manner for the future, it is surely allowable to presume on a similar continuity and gradation. We have also seen that whatever the specific phases of vitality, they never diverged beyond certain limits, but were ever constructed after a few grand types and models; and, believing in the continuity of natural law, we rely on the future adherence to the same primal patterns. In fine, progressing as nature is, the life of the Future must differ specifically from that of the Present; but, speaking in general terms, the difference cannot be more than that we have traced between the life of the successive epochs of geological history.

In reasoning on the future aspects of vitality, we must ever make allowance for the influence and operations of MAN, who comes on the present stage of geological time as a sub-creative power and new modifying agent. In the olden epochs the laws of change acted solely through the operations of purely physical agents, and what under their control took ages to accomplish may now, under the agency of man, be brought about within the scope of a single century. To the materialism and mechanism of the Past we now add the mentalism of the Present—an emanation "after God's own image," and a reasoning instrument in the hands of the Creator to effect most important changes on the vitality of the globe. The modifications brought about by man in his onward progress are already remarkable, though only the merest fraction of what they are destined to be under the influences of increasing population and higher civilisation. In his onward progress of cultivation, observe how many species of plants he destroys, and how many new varieties he creates; how by his drainage and tillage he modifies soil and climate, making new conditions, obnoxious and fatal to some races, and congenial to others; and how, in taking possession of new countries, he destroys the carnivorous and dangerous animals, and substitutes the domesticated in their stead — extirpates the indigenous flora, and plants in its place the vegetation of other regions! Mark what changes the white man has wrought, within the last few centuries, on the life of the globe, in North and

South America, in Southern Africa, in Australia, and in New Zealand, by the extirpation, the introduction, and the interchange of species! With the exception of the dingo, or problematically native dog, no placental mammal was known in Australia, which lay like a belated outlier of secondary life, at the time of its discovery by European navigators; and now most of the quadrupeds of Europe are there thriving and increasing amazingly. When we turn to the New World we find the same process on an older and larger scale. All the domestic animals of Europe, naturally unknown in America, have firmly taken root in that continent, and many of them now roam in a wild state as freely as if they had been indigenous to the country. Even the "pests and vermin" of the Old World have insensibly found their way to the New; and the New has not been slow in making reprisals on the Old by the transmission of such unwelcome settlers. In the fulfilment of this great law of natural progress, the inferior races of his own kind are also vanishing before the civilisation of the higher; and, however much our sympathies may be excited by the fact, their continuance would be only to retard that Divine scheme of advancement to which everything above, beneath, and around us has ever been incessantly tending. No scheme of benevolent enlightenment can ever avert the fate of the natives of New Zealand and Australia; no project of civilisation, however ingenious, postpone the doom of the Red Indian. As the waves of Progress have successively swept away the nationalities, pre-historic and historic, of Asia and Europe, so the same tide is irresistibly swelling towards the obliteration of mental and moral inferiority in other regions. The order has gone forth from the beginning: its execution is inevitable.

Observe, then, what an amount of extirpation, interchange, and transmission of species has been effected by man within the lapse of a few centuries ; and note how impossible it is to predicate of future life-changes where such a power has been superinduced upon the purely physical agencies of nature! It is true that man's influence has its limit. He may modify, but he cannot create—extirpate, but cannot replace—may alter the distribution, but cannot change the character of functional performance. Over and above him are the great external conditions of nature, to which he is as subject as the meanest creature he modifies ; but within certain limits he acts as a sub-creator, and this influence must ever be allowed for in all our reasonings on the future aspects of vitality.

[Progression or Succession?]

There is just one other speculation, and we can scarcely avoid adverting to it. We have traced a progress in the past, we perceive a progress going on around us, and we presume an analogous progress in the future. Whether, then, is this progress part of some great recurring succession, or is it a progression from a beginning we cannot trace to an eternity of which we cannot even imagine? Nature operates in great *successions* as well as in what appears to our limited observation a great cosmical *progression*. We note the movement of the silent shadow on the sun-dial, and were our observations limited to the space between morning and mid-day, we might fairly question whether this slowly-progressing shadow went forward and forward for ever, or whether it did not form part of a recurring succession. We watch, and the shadow attains its meridian, falls back, and again commences its progress to-day as it did on the yesterday, and as we presume it will do on the tomorrow. As the earth daily on her axis, so also in her orbit she obeys a great law of annual recurrence—a succession that might readily be mistaken for progression by one whose observations were limited to a few long days in the month of June. The magnetic needle, which in 1660 pointed due north in London, began in 1662 to diverge to the westward, till, in 1815 (a lapse 155 years), it pointed $24\tfrac{1}{4}°$ west of north. Since 1815 it has been gradually returning from this extreme divergence, and we therefore regard it as obeying some law of secular succession. So also with the polar direction of the earth's axis, which we usually regard as pointing to one spot or "fixed point" in the heavens— viz., the "Polar Star." This, however, is not strictly correct. The pole moves very slowly, so as to describe very nearly what is called *a small circle* in the heavens. This small circle, and the motion of the pole along it, are such that, in 12,000 or 13,000 years, the pole will be distant from the present pole by more than 40°; but in some 25,000 years it will have returned to the point in the heavens which it now occupies. In the geologic ages we have seen again and again the return of cold and warm influences to the same latitudes. First, the icy sterility of the Cambrian grits; second, the doubtful glaciers of the old red; next those of the Permian or new red; and again, those of the boulder-drift that immediately preceded the current era. These are also indicative of great secular recurrences in nature — successions rather than continuous progression. May it not be so with the World itself? Is it going forward, with all its physical mutations and garniture of life, from a beginning philosophy cannot trace to an end that fancy cannot dream of? Does nature never repeat herself; or is all that has taken place only part of a great succession that will again be repeated? Is there in reality nothing new under the sun, and is that which now exists only that which already has been? or is there not, as implied in the facts of geology, a recurrence of infinitesimally divergent phenomena that assume the stamp and character of an ever onward progress? From the restricted nature of individual life we are unconsciously led to associate with everything around us the idea of a beginning, a progress, and an end. An endless progression, like an eternity of unchangeable sameness, is a notion we cannot realise; and we are apt to regard the successive phases of geological history as mere stages in a progression which has had its beginning in the Past, and must come to an end in the Future. The beginning and end of this progression, however widely separated, may after all only mark the limits of a single stage in some vaster scheme of progress; and what seems to terminate the present may only be the beginning of another and higher phase of terrestrial vitality.

[Onward and Upward.]

Ignorant of the teachings of geology and the great progression it unfolds, mankind have hitherto regarded the scheme of life as culminating and terminating with their own race. All or nearly all the hopes that give colouring to their thoughts and direction to their actions proceed from this belief, though in strictest science the belief itself rests on no logical foundation. It is true, one of our highest biological authorities (Professor Agassiz) "thinks it can be shown by anatomical evidence, that man is not only the last and highest among the living beings of the present period, but that he is the last term of a series, beyond which there is no material progress possible in accordance with the plan upon which the whole animal kingdom is constructed; and that the only improvement we can look for upon earth, for the future, must consist in the development of man's intellectual and moral faculties." This, however, is a mere plausible assertion; the "anatomical evidence" is not produced; and every one cognisant of the history of man knows that intellectual and moral development has ever been restricted to the newer and advancing varieties of our race. It is true that man at present stands the crowning form of vital existence, but the facts of the past give no countenance to the belief that he shall remain the crowning form in future epochs. From its dawn until now the great evolution of life has been ever upward, geologically speaking (and be it borne in mind we are treating the question solely from a geological stand-point): shall it not continue to be upward still? We see no symptom of decay either in the physical or vital forces of nature; and so long as these forces continue to operate, mutation and progress must inevitably follow. Man's own history, physical and moral, has been one of incessant change and progress. The features of different races, their mental qualities, civil systems, and religious beliefs, have all less or more partaken of this mutation; and the difference that now subsists between the most intellectual, city-dwelling, machine-making Anglo-Saxons and the men of the old flint-implements and bone-caves may be infinitesimally small when compared with that which may exist between the noblest living nations and races yet to be evoked. Unless science has altogether misinterpreted the past, and the course of Creation as unfolded by geology be no better than a delusion, the future must transcend the present, as the present transcends that which has gone before it. Man present cannot possibly be man future. Noble as he may appear in his highest aspects, it were to limit creative power and arrest its progress to aver that man may not be superseded by another form still nobler and more divine. Physiologically, we cannot suppose that the homologies of the vertebrate skeleton have been exhausted in the structural adaptations of man : psychologically, we dare not presume against the correlation of a nobler intellect with a higher organisation. On the contrary, in these ascending forms the divine idea of moral perfection, though unconceivably unattainable by created existences, may be nearly and more nearly approached, and stage by stage the loftiest and holiest aspirations of the present may become the realisations of the future. To speculations such as these, though lying fairly in the way of geological inquiry, science can do little more than merely indicate the line of reasoning ; and if they shall be thought to involve any question as to man's religious beliefs and his hopes of a future life, on this point also science is mute, and defers with humility to the teachings of a higher philosophy.

CONCLUSION.

AND now my task is finished. I have endeavoured, in tracing the long line of Past Life, to assimilate its extinct forms to those now existing, that we may be enabled to catch a glimpse, however faint, of the unity and connection that run throughout the whole. Impossible as it was, within the limits of this Sketch, to enter into minute details, I have restricted myself to such an outline as might, with a little previous information, be intelligible to the majority of general readers, or which, in the want of that information, might be readily filled up by the perusal of any of the ordinary works on Geology. To those who may sneer at "smatterings of science," or grow facetious on the "dangers of a little learning" (and these are generally the mere technical tradesmen of some narrow department), I have only to answer, that a beginning must be made somewhere—that the little learning of to-day may form a foundation for the larger stock of the to-morrow—and that the mind is more likely to be stimulated to further inquiry by the generalisations of a vivid outline than by an array of details, the very nomenclature of which is often a puzzle and perplexity.

Whatever the amount of information conveyed, one of the main objects has been to keep prominently in view the operation of natural law, and to discourage the common but mistaken idea of the cataclysmal and revolutionary in the past history of the globe. There can be no true notion of nature or of nature's requirements so long as her facts are viewed through the medium of the miraculous or abnormal; and it were greatly to be desired that in social and moral, as well as in natural science, we should learn to recognise in every instance the fixity and unerring operation of Law, and so cease to ascribe to the blind deity of Fate what our own knowledge ought to teach us to avoid and enable us to avert. Nor let it be thought, we again repeat, that by so doing we place a wider distance between the Creator and his works, or that any knowledge of this kind has a tendency to self-sufficiency and irreverence. Law is but the mode in which the Creator has chosen to manifest himself in his works, and the highest attainment of reason is to give intelligible expression to these modes, so that we may be enabled to determine their courses and anticipate their results. For this purpose I have endeavoured, throughout the preceding review, to group and associate facts, and therefrom to deduce such generalisations as seem warranted by the teachings of Palæontology. Where the objects of research are so fragmentary and obscure, where so few of the innumerable forms entombed in the crust of the globe can have yet been exhumed, and where so little has been done in distant regions to discover and identify contemporaneous formations, I am fully aware how provisional and temporary such generalisations must necessarily be. In the mean time, however, they serve as centres round which to marshal new facts, and they give consistency to what might otherwise appear a mass of heterogeneous and not unfrequently contradictory details.

And speaking of *facts*, I would here, in the name of Palæontology, solicit that assistance which lies, less or more, in the power of every one to afford. The objects of research are scattered everywhere; and every chip and fragment that bears on it the impress of organic structure, however worthless it may appear to him who stumbles against it, may be the means not only of restoring a new form to the life of a former epoch, but the means of suggesting the connection that leads to the determination of some great creational law. Much as has been done within the last twenty years, we still stand greatly in need of additional data; and without an extensive array of facts whereon to found our generalisations, the laws that regulate the great cosmical evolution of vitality must remain, in a proportional degree, uncertain and obscure. Nor let it be thought that any devotion to palæontology—to the "stocks and stones" of the sneerers at science—will ever lessen our love for the fresh and beautiful in existing nature. To him who has traced with appreciation the long line of vegetable evolution, the flowers will bloom with new lustre, the woodlands with fresh verdure, and the solemn forest-growths inspire unwonted adoration and awe. To the student of the Past the lowest shell-fish may claim an ancestry that excites new interest; the meanest reptile may retain some curious feature of its gigantic prototypes; and some obscure and solitary quadruped may be the last of a line that once held regal sway in the forests of prehuman epochs. As the existing throws new light on the extinct, so the extinct adds fresh interest to the existing; and thus, to the palæontologist, the study of life becomes not only a more exciting pursuit, but a higher and more ennobling theme.

Besides these intellectual advantages, there are others of a moral kind that spring indirectly from the study of palæontology. There is no other science, perhaps, that tends to engender so much the feeling of community; none that connects more closely the whole of animated nature into one inseparable system. It shows that life existed before we were; it indicates that life may exist after mankind has ceased to be. Evade and resent as we may the idea of a genetic connection with the lower animals, there is no gainsaying the fact that with them we constitute part and parcel of a great vital plan. They are our life-comrades; they suffer hunger and thirst as we do; they are happy under pleasure, and miserable under pain. Exalted above them by a higher intellect and the gift of moral perception, we are bound to extend to them the humanity of our position; and we err against the Creator's scheme the moment we deal with them otherwise than is indicated by the great law of interdependence which palæontology reveals. And if we are thus led by cosmical considerations to extend mercy to our fellow-creatures, much more are we called upon to exercise it towards our fellow-men. It were a sorry account of our knowledge of the material and vital worlds, and the laws by which they are governed, did we fail to apply it to the material and moral welfare of our race. Vanity and vexation of spirit, did the tree of knowledge ripen no fairer fruit than the pride and boast of knowing! In this way the philosophy of our science ascends above the mere materialities of the earth, and becomes portion of the higher philosophy of the heart and soul.

And now, and in the last place, a word on the spirit in which we should inquire. Geology is at best but a recent science, and its task (as yet but imperfectly performed) is a very wide and difficult one: wide, as embracing a vast field of co-relative science; and difficult, as the objects of research can only be obtained by great labour, are often obscure, and, for the most part, far removed from their producing causes. In this case, though the history of the past be ever attractive, its elimination requires extensive travel and careful research. Guided solely by a desire to arrive at Truth, our observations must be made with great caution; and even with the utmost care we must often remain contented with mere description—confessing, and not ashamed to confess, that the facts observed are beyond our explanation. To observe without being biased by preconceived theory—to describe accurately so that others may reap the legitimate fruits of our observation—to advance our opinions with humility, where there is so much liability to error—and to deal charitably towards the opinions of others—are duties, without the exercise of which no man can be said to be imbued with the right spirit of geology. It has been nobly said, that "to do justly, to love mercy, and to walk humbly," are the chief requirements of moral duty: would the same spirit were ever reverently carried into matters of scientific investigation! It was for the want of these qualities that the early course of geology was so much obstructed; it is still for the neglect of their exercise that so much contention prevails, and that humble honest truth is so often over-ridden by bold-faced ignorance and dogmatism.

Guided by this spirit, and exercising it within her own proper field, a glorious future lies before geology—that future being nothing short of a perfect history of our planet. We say, exercising it *within her own proper field;* for it cannot be denied that many, assuming to themselves the character of geologists, indulge in speculations for which the science is not fairly accountable. "Theories of the Earth," "Vestiges of Creation," "Untieings of the Geological Knot," "Pre-Adamite Sketches," and "Scriptural Reconciliations," are ever crowding thick upon us— enough to destroy the reputation of any science not founded on the sure and ample bases of Truth and Philosophy. The day for a veritable theory of the World is yet far distant; let us content ourselves in the mean time by labouring diligently in the way. The vestiges of vital development are yet but faintly discernible; we shall never trace them to their source and origin under the guidance of a materialistic hypothesis. To attempt, on the other hand, reconciliations of geology with Scripture is to mistake the functions of both—to confound the philosophically ascertainable with what needed to be revealed—the physical with the spiritual, and reason with faith. It has been replied, no doubt, that the Words and the Works of God cannot possibly be at variance. This, however, is a mere dignified nothingism. No rational man ever supposed they could, but men may differ in their interpretation of either, and this makes all the difference. Geology loses by such well-meant but ignorant attempts—theology cannot be a gainer.

Let us then, as geologists, restrict ourselves to our own proper field—the physical evidences of God's working in creation, labouring to comprehend his plan, and from a comprehension of that plan to rise to the higher conception of his will as regards our own place and function in the scheme of vitality. To combine our knowledge of the earth's history as an intellectual attainment with the practical application of its treasures to our material necessities, is a high and important aim; to ascend from this aim to the conception of the whole as an orderly Cosmos, with whose ordainings, physical and vital, our thoughts and actions are inseparably interwoven, is the loftiest attainment—the true philosophy of geology. As yet this height has lain far and dimly before us; and the path of the earlier travellers has been often uncertain and obscure. Light, however, is beginning to crest the mountain-tops, and objects to cast their shadows across the valley below. Yet a little longer, and the sun will attain its meridian, and bathe in the light of knowledge all that is permissible and possible to be known. Let us take care, lest by presumptuous generalisation, by illiberality to the opinions of others, by the want of moral courage to avow the truth as it appears to us, or by giving way to unworthy prejudices, we should do aught to retard such a devoutly-to-be-wished-for consummation.

INDEX.

INDEX.

	PAGE
Acanthodes, Lower Old Red species, figured,	98
Acrogens, their aspect and function,	41
Adiantites, gigantic fern from Irish Old Red, figured,	93
Æpyornis, extinct bird of Madagascar,	164
Agassiz, Professor, quoted,	183, 201
Agassiz, on immortality of animals, quoted,	206
Agencies, nature of, affecting the globe,	71
Amblypterus, ganoid fish of coal-measures,	110
Ammonites, oolitic cephalopods, figured,	135
Amphigens, their aspect and function,	40
Ananchytes, characteristic chalk sea-urchin, figured,	145
Ancyloceras, characteristic chalk cephalopod,	147
Animal life, its governing conditions in space,	48
Animals, systematic classification of,	57
Annelid burrows, or tracks of marine worms (?),	94
Anoplotherium, restored outline of form,	156
Antholites, fossil flowers of the Carboniferous,	105
Archæocidaris, carboniferous, plates and spine of,	107
Archæogosaurus, carboniferous reptile,	110
Archæology, interest and importance of,	18
Articulata, aspect and apparent functions of,	63
Avicula, oolitic species, figured,	134
Barnacles, occurring fossil in oolite,	132
Belemnites, internal shells of fossil cuttle-fishes,	135
Bellerophon, carboniferous gasteropod, figured,	109
Bellinurus, limuloid carboniferous crustacean, figured,	108
Beryx, fish of cretaceous epoch, figured,	148
Birds, occurrence of in cretaceous strata,	149
Birds, supposed first appearance of,	127
Boreal shells from valley of the Clyde,	167
Botanical arrangement, principles regulating,	33

INDEX.

	PAGE
Bothrodendron, pitted stem of the Carboniferous,	104
Boulder-clay or Glacial period,	165
Boulders or drift-masses imbedded in chalk,	143
Bramatherium, gigantic tertiary mammal from India.	159
Cainozoic or "Recent Life" systems of strata,	151
Calamites, reed-like stems of the coal-measures,	104
Calceola (a little slipper), Devonian shell, figured,	94
Cambrian era, its vital characteristics,	82
Carboniferous era, its physical and vital characteristics,	100
Carboniferous flora, its peculiar aspect and nature,	102
Carpolites, fossil fruits of the Carboniferous,	105
Caulopteris, tree-fern stem of the coal-measures,	104
Cell, the vegetable, nature of,	33
Centres of creation, for human species,	216
Cephalaspis, characteristic fish of Lower Old Red,	97
Cephalopods of the chalk, curious configurations of,	147
Cephalopods of the Oolitic era, various, figured,	135
Ceratiocaris, palæozoic bivalved crustacean,	89
Cereals, their first appearance,	152
Chondrites, Silurian sea-weed, figured,	85
Conclusion, general review of generalisations,	241
Climate, its influence on plant life,	29
Climatius, spiny-fish of Lower Old Red,	98
Clisiophyllum, carboniferous coral, figured,	106
Clymenia, characteristic Devonian shell, figured,	94
Coal-fields of the Oolitic era,	132
Coccosteus, characteristic fish of Lower Middle Old Red,	97
Colder and warmer cycles, hypothesis of,	191
Coniferous trees of the oolite, seasonal growth,	130
Connecticut Valley, fossil footprints of,	126
Conularia, carboniferous pteropod, figured,	109
Coprolites, or fossil droppings, carboniferous,	109
Corals, various carboniferous genera, figured,	106
Corals, various Silurian forms, figured,	87
Co-relation of parts, Cuvier's law of,	53
Creation, apparent course of, as indicated by Geology,	221
Creation still in progress,	226
Creations, new, how and by what means effected,	194
Cretaceous era, physical and vital features of,	142
Crioceras, characteristic cephalopod of chalk,	147
Crust of the earth, its composition and structure,	69
Crustacea of oolite, their characteristics,	133
Cruziana, Silurian sea-weed, figured,	85
Ctenacanthus, carboniferous fin-spine, figured,	111
Ctenoptychius, palatal tooth of carboniferous fish,	111
Current or Human era, its aspects,	168
Cuttle-fishes, or naked cephalopods of oolite.	135

INDEX. 251

	PAGE
Cyathophyllum, carboniferous cup-coral, figured,	106
Cycadaceous plants of the Oolitic epoch,	132
Cypris, small bivalved crustacean of coal,	108
Darwin, on origin of life, quoted,	210
Darwin's hypothesis as to origin of species,	197
Death, consideration of, in scheme of vitality,	185
Dendrerpeton, lizard-like reptile of Coal era,	110
Development hypotheses, considered,	197
Devonian flora, various fragments of, figured,	92, 93
Devonian, or Old Red Sandstone era,	91
Dinornis, great fossil bird of New Zealand,	164
Diplacanthus, Lower Old Red species, figured,	98
Diprotodon, extinct kangaroo of Australia,	159, 166
Distribution of life never uniform,	186
Dithyrocaris, carboniferous bivalved crustacean,	108
Dodo, extinct bird of the Mauritius,	164
Dromatherium, jaw of, from North America,	115
Egg-packets, supposed fossil spawn of crustacea,	96
Electricity and vital phenomena,	181
Elgin sandstones, containing reptilian remains,	99
Embryology, nature of the doctrine of,	202
Encrinites, Silurian genera, figured,	87
Encrinites, various carboniferous forms, figured,	107
Endogens, their aspect and function,	44
Eocene flora and fauna, their characteristics,	155
Equisetites, fossil stems resembling the equisetum,	104
Euomphalus, carboniferous gasteropod, figured,	109
Eurypteridæ, palæozoic crustacea, characters of,	95
Eurypterites, Silurian crustacea, figured,	89
Eurypterus or Idothea, carboniferous crustacean, figured,	108
Exogens, their aspect and function,	45
External conditions of life never uniform,	188
Extinction or creation of species, never general,	195
Fauna and flora, their mutual co-adaptations,	66
Flint implements from upper pleistocene, figured,	170
Flora and fauna of the Current epoch,	168
Flora and fauna, their mutual co-adaptations,	66
Footprints, how occurring in a fossil state,	126
Foraminiferæ, their abundance in chalk rocks,	143
Fossils, interest attached to,	18
Fossils, nature and character of,	20
Fresh-water formations, occurrence of,	154
Fresh-water shells of the Tertiary era,	157
Function, uniformity of, in all time past,	184
Functional adaptations of animal life,	51

	PAGE
Future life-periods, probable aspects of,	174
Future vitality, man's influence on,	232
Galerites, characteristic chalk sea-urchin, figured,	145
Generation, spontaneous, unsatisfactory evidence of,	181
Geological classification—systems and periods,	73
Geology, her own proper field of inquiry,	245
Geology, spirit in which to study,	244
Geology, the vital problems yet to solve,	75
Glacial action, appearances of, during Old Red era,	91
Glacial or Northern Drift epoch,	165
Glyptodon, extinct gigantic armadillo,	163
Goniaster (corner-star), chalk star-fish, figured,	145
Goniatite, carboniferous cephalopod, figured,	109
Graptolites, Silurian zoophytes, figured,	86
Gryphæa, characteristic shell of the Oolite,	134
Gymnogens, their aspect and function,	42
Gyracanthus, carboniferous fin-spine, figured,	111
Hamites, characteristic chalk cephalopod,	147
Hemipneustes, sea-urchin of Cretaceous epoch, figured,	145
Higher and lower, as applied to organised beings,	55
Hippurites, characteristic cretaceous bivalve,	146
Homology, the anatomical doctrine of,	53
Human race, mutability of,	173
Human race, probable antiquity of,	215
Human race, probable duration of,	230
Hylæosaurus, great land saurian of the Weald,	138
Hypotheses, their value in science,	178
Hypozoic or metamorphic strata,	81
Ice-action, appearances of, during Devonian period,	91
Ice-action, supposed occurrence of, in Permian,	115
Ichnites, or fossil footprints, how occurring,	126
Ichnology, or science of fossil footprints,	126
Ichthyosaurus, oolitic marine reptile, figured,	137
Igneous or fire-formed rocks, defined,	69
Immortality of life-forms, question of,	206
Implements, flint, what they indicate,	217
Inoceramus, cretaceous species of, figured,	146
Inquiry, spirit of, necessary for geology,	244
Insects, various genera, fossil in oolite,	133
Irish gigantic deer, sketch of, figured,	169
Labyrinthodon, lower secondary reptile, figured,	125
Lamarckian hypothesis of vital development,	197
Law, natural, its nature and importance,	242
Law of similar habit and economy,	199

INDEX. 253

	PAGE
Law, uniformity and universality of action,	80
Laws apparently regulating past vitality,	177
Leaf, as the primary structural organ,	32
Lepidodendron, characteristic plant of the Carboniferous,	104
Life, dawn of, as known to geology,	178
Life, origin of, wholly unknown,	180
Life-periods of geology, arrangement of,	73
Lignites and coals of the Cretaceous epoch,	143
Lingula, Silurian brachiopod, figured,	88
Lituites, Silurian cephalopod, figured,	88
London basin, its flora and fauna,	155
Loxonema, carboniferous gasteropod, figured,	109
Machairodus, fossil cave-lion of Europe,	159
Maclurea, characteristic Silurian shell, figured,	88
Mammals in Permian sandstones of America,	165
Mammoth, or extinct hairy elephant,	162
Man, his advent or first appearance,	170, 213
Man, his advent in accordance with the general plan of vitality,	215
Man, his early condition, as indicated by geology,	217
Man, his influence on future life-aspects,	232
Man, inferior races of, their inevitable decay,	234
Man, not the last of this vital series,	237
Marsupial quadrupeds of Triassic era,	128
Marsupials of the Oolite, jaws of various,	139
Marsupites, cretaceous echinoderms, figured,	144
Mastodon, extinct elephantoid mammal,	162
Megaceros Hibernicus, or great Irish deer,	169
Megalanea, extinct lace-lizard of Australia,	159
Megalodon (large toothed hinge), Devonian shell, figured,	94
Megalosaurus, great land saurian, figured,	138
Megatherium, extinct gigantic ground-sloth,	163
Merycotherium, fossil mammal from India,	159
Mesozoic systems, their chief characteristics,	119
Microlestes, insectivorous mammal of trias,	128
Migration and migratory races at all periods,	187
Modiola, oolitic bivalve, figured,	134
Mollusca, aspect and apparent functions of,	62
Mollusca of the Oolite, various genera, figured,	134
Monkey, development of, considered,	205
Moral perception, restricted to man,	205
Murchisonia, gasteropod, carboniferous species of,	109
Murchisonia, gasteropod, Devonian species, figured,	94
Murchisonia, gasteropod, Silurian, figured,	88
Natural selection, Mr Darwin's theory of,	204
New life forms, how introduced,	192
Nummulites and nummulitic strata,	157

	PAGE
Oldhamia, Silurian polyzoan, figured,	86
Old Red Sandstone or Devonian era,	91
Oolitic era, physical and vital features of,	128
Oolitic era, physical geography of,	141
Oolitic flora, restored aspect of,	130
Orbitoidal limestone of North America,	158
Orbitoides, abundant foraminiferous forms,	144
Orthoceratite, carboniferous species, figured,	109
Orthoceratite, Silurian, figured,	88
Osmeroides, fish of chalk formation, figured,	148
Pain, consideration of, in scheme of life,	185
Palæchinus, carboniferous sea-urchin, figured,	107
Palæoniscus, ganoid fish of Carboniferous era,	110
Palæoniscus, Permian species, figured,	114
Palæontology, advantages of the study of,	242
Palæontology, its general scope and function,	25
Palæontology, its intellectual bearings,	24
Palæontology, its practical bearings,	33
Palæontology, science of ancient beings, defined,	20
Palæophytology, or science of fossil plants,	46
Palæotherium, restored outline of form,	156
Palæozoic, or ancient life-systems,	79
Palæozoology, or science of extinct life,	65
Palapteryx, or extinct apteryx of New Zealand,	159, 166
Palms, doubtfully occurring in the coal-measures,	103
Pampas of South America, their fossils,	163
Permian, or Lower New Red Sandstone era,	114
Phascolotherium, jaw of, from the upper Oolite,	140
Phillipsia (after Professor Phillips), carboniferous, trilobite,	108
Plagiaulax, jaw and teeth from the upper Oolite,	140
Plan and order of life pre-ordained,	184
Plants, systematic classifications of,	37
Platysomus, Permian fish, figured,	114
Plesiosaurus, Oolitic marine reptile, figured,	137
Pleuracanthus, carboniferous fin-spine, figured,	111
Pleurotomaria, carboniferous species, figured,	109
Pœcilodus, carboniferous palatal tooth of fish,	111
Polyzoa of the chalk, their characteristics,	145
Post-tertiary and Tertiary, definitions of,	153
Present, the, its flora and fauna,	27
Productus, characteristic carboniferous brachiopod,	109
Progression or succession in nature,	235
Protozoa, aspect and apparent functions of,	59
Provinces of animal life,	50
Psammodus, carboniferous palatal tooth of fish,	111
Psiloplyton, plant from Canadian Old Red, figured,	92
Pterichthys, characteristic fish of Middle Old Red,	97

INDEX. 255

	PAGE
Pterodactyle, flying reptile of the Oolite, figured,	138
Pterozamites, pinnate leaf of, Triassic era,	125
Pterygotus anglicus, from Lower Old Red, figured,	95
Pterygotus, upper Silurian species, figured,	89
Radiata, aspect and apparent functions of,	60
Radiolites, cretaceous bivalve, figure of,	146
Rain-prints, or impressions of ancient showers,	94
Ramsay, Professor, on Permian glaciers,	115
Representative species, what is meant by,	50
Reptiles, carboniferous, lowly nature of,	110
Reptiles, gigantic secondary, their functions,	185
Reptiles of the Old Red, still doubtful,	99
Rhizodus, sauroid fish, jaw and dentition of,	111
Rock-systems of geology, arrangement of,	73
Rotalia, cretaceous foraminiferous organism,	144
Rytina, extinct dugong of Kamtschatka,	172
Sauroid fishes of Carboniferous era,	110
Scaphites, characteristic chalk cephalopod,	147
Scriptural reconciliations, inutility of,	245
Seals, sub-fossil, in pleistocene strata,	167
Sedimentary, or water-deposited rocks, defined,	70
Sigillariæ, gigantic trees of the Carboniferous era,	104
Silurian era, physical and vital characteristics,	83
Silurian flora, scanty and fragmentary,	84
Siphonia, silicified spongiform organ in chalk,	144
Sivatherium, sub-fossil ruminant of India,	159
Solitaire, extinct bird of Rodriguez,	172
Species, duration of, in time,	228
Species representative and identical,	50
Spirifer, carboniferous species, figured,	109
Spirorbes, calcareous tubes of, in coal-measures,	106, 108
Sponges and their function in cretaceous strata,	143
Star-fish, Silurian, figured,	87
Stigmaria, roots of the carboniferous sigillaria,	104
Stone implements of the upper pleistocene,	170
Stringocephalus, characteristic Devonian shell, figured,	94
Strophomena, Silurian brachiopod, figure of,	88
Stylonurus, crustacean from Forfarshire Old Red,	95
Succession or progression in natural events,	235
Superposition of strata, key to geological time,	70
Telerpeton, lacertilian reptile from Elgin sandstones,	126
Temperature, internal, its effect on climate,	189
Terebratula, carboniferous species, figured,	109
Tertiary and post-tertiary strata, defined,	153
Tertiary, its different chronological stages,	157

	PAGE
Textularia, abundant foraminiferous organism,	144
The Record—as depending on Geology,	69
Thecodont reptiles, their high position,	199
Theory of the World, Geology not prepared for,	245
Time, its estimate in years and centuries,	220
Time, pre-geological and geological,	218
Transitional or intermediate forms,	198
Triassic era, physical and vital conditions of,	121
Triconodon, jaw and teeth of, from Upper Oolite,	140
Trigonia, Oolitic bivalve, figured,	134
Trigonocarpon, supposed coniferous fruit,	105
Trilobites, characteristic Silurian crustacea, figured,	88
Trogontherium, gigantic fossil beaver of Europe,	160
Turrilites, characteristic chalk cephalopod,	147
Type and pattern, uniformity of, in life,	182
Typical order of nature,	31
Uniformity of natural law,	71
Varieties, their limit and continuance,	202
Vegetation, the existing zones of,	28
Vegetation, effects of altitude on,	29
Vegetation, existing provinces of,	30
Ventriculites, fossil chalk sponge, figured,	144
Vertebrata, aspect and apparent functions of,	64
Vestiges of Creation, its materialistic views,	197
Vestiges of Creation, referred to,	209
Vital hypotheses, how to receive,	208
Walchia, triassic coniferous stem, figured,	125
Wellingtonias of California, limited range of,	173
Xiphodon, restored outline of form,	156
Zoological arrangement, principles of,	52
Zones of depth, as influencing marine life,	49
Zosterites, Devonian sea-weed, figured,	92

THE END.

PRINTED BY WILLIAM BLACKWOOD AND SONS, EDINBURGH.

www.ingramcontent.com/pod-product-compliance
Lightning Source LLC
Chambersburg PA
CBHW021401230426
43666CB00006B/598